SpringerBriefs in Computer Science

More information about this series at http://www.springer.com/series/10028

Beniamino Di Martino · Giuseppina Cretella
Antonio Esposito

Cloud Portability and Interoperability

Issues and Current Trends

Beniamino Di Martino
Seconda Università di Napoli
Aversa
Italy

Antonio Esposito
Seconda Università di Napoli
Aversa
Italy

Giuseppina Cretella
Seconda Università di Napoli
Aversa
Italy

ISSN 2191-5768   ISSN 2191-5776   (electronic)
SpringerBriefs in Computer Science
ISBN 978-3-319-13700-1   ISBN 978-3-319-13701-8   (eBook)
DOI 10.1007/978-3-319-13701-8

Library of Congress Control Number: 2014956494

Springer Cham Heidelberg New York Dordrecht London

Springer International Publishing AG Switzerland is part of Springer Science+Business Media (www.springer.com)

To Anna, Francesco, Rosalba
Luca, Teresa, Virgilio, Francesca, Gio
Paola, Rosaria, Francesco

Preface

Cloud computing is emerging as the most promising technology for software development, changing the way customers interact with their data and applications. There are many reasons that drive the choice of moving to cloud:

- companies no longer need to buy, store, and maintain expensive hardware infrastructures, reducing time and money involved in maintaining, updating, and repairing their own equipments;
- hardware dimensioning is not to be related to peak workload any more, but the infrastructure can be dynamically scaled according to the current needs. This results in a better use of the existing resources;
- customers pay only for the resources they actually use, following a "Pay as you Go" paradigm;
- using distributed resources, including data centers and computing nodes, can enhance systems' resiliency and disaster recovery;
- the possibility to choose among a broad range of available resources and services can trigger strong competition between cloud providers, thus resulting in better quality and lower prices for customers.

These are only a few of the possible benefits that could derive from the adoption of the cloud computing paradigm but, despite the diffusion of cloud technologies, issues and limitations still exist. A major issue is the lack of portability and interoperability between cloud platforms at different service levels, affecting the cloud computing panorama in several ways and aspects. The brokering, negotiation, management, monitoring, and reconfiguration of cloud resources are nowadays challenging tasks for the developer or user of cloud applications due to different business models associated with resource consumption as well as due to the variety of services—and their features—offered by the variety of cloud providers. These points become very critical when the landscape is a multicloud environment and the main concern is represented by the *vendor lock-in* problem. In fact, cloud providers usually propose technological solutions that differentiate them from their competitors: these differences have the drawback of locking the customers as no alternatives are offered. Thus, once customers have chosen a cloud provider, either they cannot change to another provider or they can do it but only at a huge cost. Vendor lock-in risk also includes reduced negotiation power in reaction to price increases and service discontinuation (if, e.g., the provider goes out of business).

In the following the structure of the book is illustrated.

In Chap. 1, the notions of cloud portability and interoperability are introduced, together with the issues and limitations arising when such features are lacking or ignored. The illustration starts with definitions of portability and interoperability as inherent to the generality of software systems, and then the concepts are tailored, specialized, and exemplified for the specificity of cloud computing. Some basic concepts of the cloud and reference architectures are reported to define the recurrent terms and roles.

A number of use cases, accompanied with a concrete case study, representing a variety of interoperability and portability scenarios are illustrated. Several definitions and use case scenarios are modeled by means of an n-dimensional feature space, where features represent the different characteristics and abstraction levels of the cloud domain.

The feature space, the use case scenarios, and the case study are utilized in the following chapters in order to position, classify, and demonstrate the different technologies and solutions presented.

Chapter 2 provides an overview of the state-of-the-art methodologies and technologies, which are currently used or are being investigated to enable cloud portability and interoperability. These include: Model-Driven Architecture (MDA) and languages, semantic technologies, cloud patterns, and agent systems. We illustrate in detail how the use of cloud patterns can enable robust cloud applications design and development with respect to portability and interoperability. We position the different methodologies and technologies illustrated with respect to the use case scenarios and features defined in Chap. 1, and we test them by analyzing their application to the case study illustrated in Chap. 1. We also mention and briefly illustrate the contributions coming from projects funded by the European Commission FP7 program.

Chapter 3 illustrates the main cross-platform cloud application programming interfaces and how they can solve interoperability and portability issues by bringing uniformity and standardization to the cloud. This chapter provides an overview of initiatives that provide cross-platform-based cloud APIs such as DeltaCloud, SimpleCloud, JCloud, Libcloud, and research projects whose aim is to provide multicloud APIs (such as mOSAIC). Such APIs are positioned with respect to the use case scenarios and features defined in Chap. 1, and tested by analyzing their application to the case study illustrated in the chapter.

Chapter 4 presents a set of ready-to-go solutions which, either for their wide diffusion in the cloud computing scenario or because they implement the established or emerging standards (see Chap. 5), have a fundamental role in providing interoperable and portable solutions. In particular, Amazon Web Services (AWS) and OpenStack have imposed themselves as "de facto standards" at the IaaS level, since their wide adoption has led other providers to develop APIs and interfaces which are compatible with their offers. At the PaaS level, Microsoft Azure, Google App Engine, IBM Bluemix, and OpenShift with their multi-language support and ability to interface with other platform services, surely enhance the application portability and thus deserve to be cited here. Such solutions are positioned with respect to the use case scenarios and features defined in Chap. 1, and tested by analyzing their application to the case study illustrated in the chapter. Finally, we also present solutions that have been explicitly created for portability and inter-operability purposes, such as Docker, ElasticBox, and Cloudify.

Chapter 5 presents an overview of the emerging standards for cloud interoperability and portability. In particular, here we consider efforts moving toward the definition of shared standards addressing different aspects of the cloud environment, spanning from services communication to data description. Among these, we consider standards such as TOSCA, CIMI, OCCI, and CDMI. Some of these standards are positioned with respect to the use case scenarios and features defined in Chap. 1, and tested by analyzing their application to the case study illustrated in the chapter.

We wish to thank Antonio Argenziano, Andrea Barbato, Graziella Carta, Salvatore D'Angelo, Salvatore Maisto, Stefania Nacchia, and Raffaele Sperandeo for their valuable contribution to the evaluation of the various platforms and solutions described in the book.

Last but not least, we would like to acknowledge the excellent and relentless support of the Springer staff members—in particular Viktoria Meyer, Ralf Gerstner, and Aliaksandr Birukou—during all phases of the development of this book.

Beniamino Di Martino
Giuseppina Cretella
Antonio Esposito

Contents

Chapter 1
Cloud Portability and Interoperability

1.1 Cloud Basics and Reference Architectures

The National Institute of Standards and Technology (NIST) has provided an accurate definition of cloud computing, which is described as "a model for enabling ubiquitous, convenient, on-demand network access to a shared pool of configurable computing resources (e.g., networks, servers, storage, applications, and services) that can be rapidly provisioned and released with minimal management effort or service provider interaction" [1]. The NIST definition, summarized in the reference architecture reported in Fig. 1.1, describes the roles within the cloud computing scenario and, for each of them, a set of corresponding capabilities and responsibilities. The five roles identified by NIST are as follows:

- **Cloud Consumer** represents either a person or an organization using the services offered by a cloud provider, generally via an online catalog, and requesting the desired services. **Service level agreements** (SLAs) specify the requirements fulfilled by a certain service or a set of services, in order to let consumers choose the appropriate ones.
- **Cloud Provider** represents either a person or an organization offering services to interested consumers. It is the cloud provider's responsibility to assure the availability of the services and the fulfillment of SLAs. One of the activity areas in which a cloud provider is involved concerns **Service Deployment**. In particular, NIST has defined the following four deployment models:

  - **Public Cloud** describes a situation in which resources and infrastructures are made publicly available over a public network. A public cloud is generally owned by a provider that sells its resources to a heterogeneous pool of consumers.
  - **Private Cloud** restricts access to services and resources to a single cloud consumer. The cloud infrastructure can be either owned and managed by the cloud consumer directly or by a third party.
  - **Community Cloud** is similar to a private cloud, since it limits access to resources that are not publicly available. However, it can serve multiple cloud consumers, sharing interests and/or objectives, rather than a single organization.

© The Author(s) 2015
B. Di Martino et al., *Cloud Portability and Interoperability*,
SpringerBriefs in Computer Science, DOI 10.1007/978-3-319-13701-8_1

- **Hybrid Cloud** represents a situation in which two or more cloud infrastructures, following different deployment models, collaborate to provide complex services. Such a composition of cloud services requires the use of standard or proprietary technologies to enable data and application portability and interoperability.

A provider is in charge of **Service Orchestration**, which is *"the composition of system components to support the Cloud Providers activities in arrangement, coordination and management of computing resources in order to provide cloud services to Cloud Consumers"*. Particularly interesting are the three service delivery models identified by NIST as part of cloud orchestration, in the **Service Layer**:

- **Infrastructure as a Service** (IaaS) is a model in which IT infrastructures, ranging from CPU power to storage, are exposed as a resource over the Internet. Cloud users can dynamically shape their infrastructure according to their needs, while resources are provided "on demand".
- **Platform as a Service** (PaaS) consists of application development platforms, remotely accessible through the web and able to connect to locally executed frameworks and IDEs, allowing fast development and deployment of applications.
- **Software as a Service** (SaaS) allows providers to expose stand-alone applications, running on a distributed cloud infrastructure completely hidden from customers, as resources accessible through the Internet.

Service levels and deployment models are represented in the following as dimensions of an n-dimensional features space in order for us to classify, position, and visually characterize portability and interoperability definitions, use case scenarios, and technological solutions.

Another responsibility of a provider is represented by **Cloud Service Management**, comprehending all the functionalities needed to correctly manage and operate services offered to consumers. In this context, a provider should also offer mechanisms to support data portability, service interoperability, and system portability (more on these will be discussed in Sect. 1.2).

- **Cloud Auditor** performs independent examinations of cloud service controls, in order to verify conformance to standards or to evaluate the provider in terms of security, privacy, performance, fulfillments of SLAs, and so on.
- **Cloud Broker** manages service negotiations and relationships between cloud consumers and providers, acting as an intermediary. It deals with request, performance tuning, and delivery of services. In particular, the services provided by a cloud broker, as they are defined by NIST, can be divided into three categories mentioned below:

- **Service Intermediation**: the broker adds value to a given service by providing some functional improvements useful to consumers, such as identity management, performance reporting, enhanced security, and so on.

- **Service Aggregation**: the broker integrates multiple existing services in order to deliver new services or functionalities. Data integration and security during transfers across multiple providers is assured by the broker itself.
- **Service Arbitrage**: the broker aggregates services from multiple providers, selecting the delivering agency according to a previously assigned score. However, the selected services are not fixed, which are presented to the consumers "as they are". In this respect, it differs from Service Aggregation in which the broker provides service integration as well.

- **Cloud Carrier** provides connectivity and transport services, enabling consumers to access the selected services through different communication devices, generally represented by the Internet.

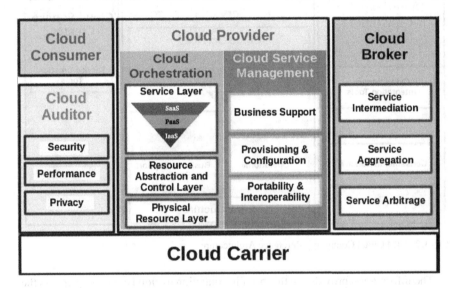

Fig. 1.1 NIST reference architecture

The NIST reference architecture represents a good starting point to understand and learn about the cloud, since many providers have been inspired by it when defining their solutions. For instance, the **IBM Cloud Computing Reference Architecture** (CCRA) [2], reported in Fig. 1.2, has been clearly influenced by the NIST model. However, IBM being very business oriented, there exist important differences between the CCRA and the NIST architecture.

- The role of **Cloud Consumers** slightly changes: while in the NIST architecture, a consumer just uses the services offered by a provider, optionally managed and organized through a broker, in CCRA she has more control over the consumed services with the possibility to integrate them with the existing **In-house IT**.
- Cloud providers can deliver their services following the three models defined by NIST (IaaS, PaaS, and SaaS), or the **Business Process as a Service** (BPaaS)

paradigm, defined by IBM itself. BPaaS consists in the delivery of business processes through cloud technologies, accessed via Internet-based technologies.

- Management tasks are supported by the **Common Cloud Management Platform** (CCMP), used to expose both operational and business support services, which are exploited by Cloud Service Creators to develop their applications.
- **Cloud Service Creator** is a new role defined in the IBM CCRA. A service creator has the responsibility to design, implement, and maintain runtime and management artifacts needed to run a specific service, which will then be exposed by providers and used by consumers. Under this perspective, the service provider and creator roles can be played by different actors, as long as the interfaces they use (exposed by the CCMP) are compliant.

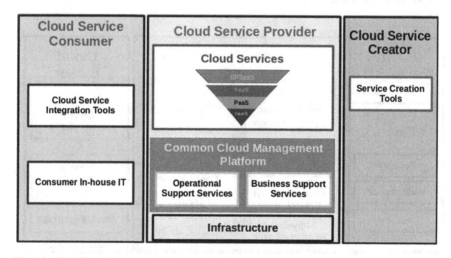

Fig. 1.2 IBM Cloud Computing Reference Architecture

The information provided in this brief introduction to cloud computing covers the basic notions needed to understand the following sections and chapters. See [3] for an in-depth and more complete discussion of cloud computing and related features.

1.2 Cloud Interoperability and Portability Definitions

Interoperability and portability are both highly desirable qualities that affect the cloud under different perspectives. Portability and interoperability, alongside the entities such features are referred to (data, services, applications and systems), represent two other dimensions of our n-dimensional features space. The n-dimensional space (Fig. 1.3) is composed of four dimensions: the software entities (data, service, application, and system), the service level (non-cloud service level, IaaS, PaaS, and SaaS), the deployment model (public, private, and hybrid), and the cloud features (portability and interoperability). A single comprehensive definition of either interoperability or portability would be too general and would not provide much information. Instead,

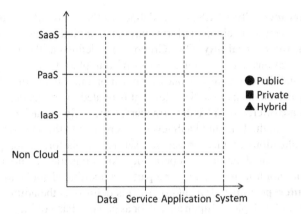

Fig. 1.3 Interoperability and portability dimensions

independent groups such as the already mentioned NIST or the OpenGroup [4] consortium have provided definitions for interoperability and portability in different cloud domains.

- **Data Portability** (NIST/OpenGroup-DP) represents the possibility, for consumers, to transfer or copy data objects to or from different cloud platforms. This property is enabled by the use of common data formats and transfer protocols, together with shared API's interfaces for data migration. Porting data can be a difficult task, since storage models and formats often vary between platforms. Also, we should consider the effort needed to move data between cloud platforms, which is not free of charge.
- **System Portability** (NIST-SP) represents the possibility to migrate virtual machine instances, machine images, applications or even services, and their relative contents from one cloud provider to another. Since it also considers applications, even if not a main focus, this definition partially overlaps that of application portability proposed by the OpenGroup.
- **Application Portability** (OpenGroup-AP) enables the reuse and migration of entire applications, or of some of their components, across cloud PaaS services or even from on-premise environments to the cloud. Such a definition contemplates two additional features:

  - The first refers to **Portability between Development and Operational Environments**. PaaS is particularly attractive because it avoids the need for investment in systems that will be dismissed once the development is complete. But, if a different environment will be used at runtime, it is essential that the applications are moved unchanged, or at least seamlessly among such environments.
  - The second refers to **Software Modernization**, which is still a significant challenge in general and even more ambitious when a change in the software delivery paradigm needs to be addressed, such as in the case of cloud computing.

- **Service Interoperability** (NIST-SI) is defined as the ability of consumers to use services across multiple cloud platforms through a unified management interface.
- **Application Interoperability** (OpenGroup-AI) is defined as the ability of cloud-enabled applications to collaborate, across different platforms, in order to deliver their functionalities or create new ones. Application interoperability affects all the service delivery models: at the SaaS level, it is related to application components or full software offered by different providers, which rely on each other to deliver functionalities; at the IaaS and PaaS levels, it is related to the specific services the involved applications require. In general, service or platform interoperability is needed to enable application interoperability. An application component may be a complete monolithic application, or a part of a distributed application.
- **Platform Interoperability** (OpenGroup-PI) is defined as the ability of platform components, either deployed upon an IaaS or as part of a PaaS offer, to interoperate. If we consider components whose functionalities are exposed as services to cloud consumers, platform and service interoperability definitions tend to coincide.

Using two of the defined dimensions (portability/interoperability vs. data/service/application/system) we are able to position the above illustrated definitions in Fig. 1.4.

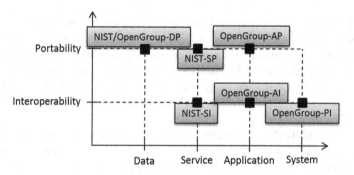

Fig. 1.4 Positioning of portability and interoperability definitions

1.2.1 Cloud Computing Use Case Scenarios

Due to the high number of variables that come into play in a complex cloud computing solution that involves portability and interoperability capabilities, a number of use case scenarios have been defined in the literature to underline the requirements and consideration of the particular case. Among the several cloud computing use case scenarios we report and classify, against our n-dimensional space, some notable examples that highlight the key aspects of cloud computing interoperability and portability characteristics. The results of our classification are illustrated in Fig. 1.5, where arrows represent interoperability/portability among different service levels, while squares represent interoperability/portability within the same service level. All the technologies, methodologies, ready-to-go solutions, and standards presented in the rest of the book will be placed with respect to such use cases. The use cases labeled with the acronym "CSCC" are presented in the document *Interoperability and Portability for Cloud Computing: A Guide* [5] published by the Cloud Standards Customer Council. The use cases labeled with the acronym "CCUS" have been defined by the Cloud Computing Use Case Discussion Group [6].

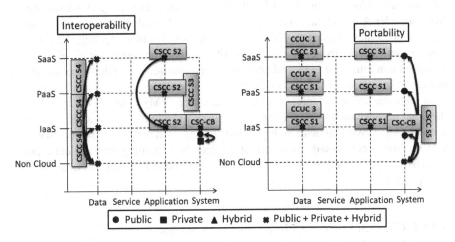

Fig. 1.5 Use case scenario classification

1.2.1.1 CSCC S1: Customer Switches Providers for a Cloud Service

This scenario addresses the case of a customer currently using a cloud service provided by provider A, who wishes to switch to an equivalent service from provider B. This scenario touches many of the issues associated with portability. From the point of view of application code, the portability strategies depend on the cloud service level: in the case of SaaS, the application code typically belongs to and is managed by the provider; in the case of PaaS, portability depends on the programming languages supported by the involved platforms and on the APIs offered by the cloud platform services to manage applications (submit the app code, configure, run, and control the app). Data portability aspects must also be considered. At the IaaS level, storage functionalities provided by cloud vendors are of typically low level, such as providing volumes for binary files or object storages, so customers are generally free to use their preferred data format. While the same sometimes occurs at the PaaS level, the situation tends to be more complex: a PaaS service provider may offer ready-to-go instances of databases, which may be sensitive to the data format chosen by the customer. However, there are some very generalized formats (CSV, XML, etc.)

that are supported by many types of databases. For SaaS services, data formats and contents are handled by the service provider, so major data portability considerations are needed.

1.2.1.2 CSCC 2: Customer Uses Cloud Services from Multiple Providers

This scenario addresses the case of a cloud service customer using cloud service 1 from provider A and cloud service 2 from provider B, while requiring to use them together to achieve business goals. Despite the benefits of mitigating the risk of data loss and temporary service unavailability, working with multiple cloud service providers can introduce logistical problems. Open-source or vendor-agnostic tools and cloud management services can help solve these issues. Not all cloud providers offer the same services. Storage and virtual machine (VM) instances are often standard, but services such as messaging or workflow and administration tools may vary across vendors. Working with multiple cloud providers may force customers to work with the lowest common denominator of services. Sidestepping this problem is possible by choosing vendor-agnostic software applications that will run in all of the target clouds.

1.2.1.3 CSCC 3: Customer Links One Cloud Service to Another Cloud Service

This scenario addresses the case of an architecture linking cloud services together to support a single application or an integrated set of applications. The advantages of multicloud deployments are several, first of all the ability of enterprises to leverage cloud solutions best fitting their stated needs, remaining within the imposed cost limits. An example of such a linking is represented by the case of a SaaS application capable of delivering basic business-related functionalities as needed by a customer, but it is not able to provide advanced functionalities (i.e., data analytics or business intelligence related). The customer can leverage the IaaS capabilities from another cloud service provider by migrating the data from the SaaS to the IaaS solutions, also in order to combine different data sets and perform more advanced analytics.

1.2.1.4 CSCC 4: Customer Links In-house Capabilities with Cloud Services

As more enterprises are planning their cloud investment, they will realize how to leverage their existing in-house IT with their future cloud setting. A proper analysis of the available APIs of both the in-house and cloud services will be required to understand how the integrated system will function and perform during typical execution.

1.2.1.5   CSCC 5: Migration of Customer Capabilities into Cloud Services

This scenario addresses the case of a customer, currently running an application or service on-premise, who wants to move that capability to a public cloud environment. For SaaS cloud services, migrating an on-premise application or service to a public cloud service provider does not involve porting the application code, because the application is being replaced. What is important in the SaaS case is the compatibility of the functional interface of the application, of any interfaces presented to end users and any APIs made available to other customer applications. In order to reduce undesired side effects, the APIs made available by the SaaS service should be interoperable with the interface provided by the on-premise application or the service that is being replaced. If the APIs are not interoperable any customer applications using the APIs will need to be changed as part of the migration process. To migrate on a PaaS cloud, the application must be designed for one of the runtime environments available in the target PaaS service. Generally a PaaS solution provides the elements of the particular software stack required by applications such as the operating system, an application server, and a database, so that the customer only has to be concerned with the specific application components and data. Some concerns may arise regarding particular configurations required by the application, such as the ability to run scripts and the presence of certain tools for setup, reporting, or monitoring. At the IaaS level, the entire software stack is migrated through one or more virtual machine (VM) images, which can then be copied into the cloud service and executed there. Some concerns arise if the application makes use of specialized device drivers or hardware devices that are unlikely to be supported by an IaaS provider.

1.2.1.6 CCUC 1: Changing SaaS Vendors

This scenario addresses the case in which a cloud customer changes SaaS vendors. The data handled by one vendor's software should be importable by the second vendor's software, which means that both applications need to support common formats. Standard APIs for different application types will also be required.

1.2.1.7 CCUC 2: Changing Middleware Vendors

This scenario addresses the case in which a cloud customer changes cloud middleware vendors. Existing data, queries, message queues, and applications must be exportable from one vendor and importable by the other. The requirement to achieve this porting is a common API for cloud middleware. Cloud database vendors have enforced certain restrictions to make their products more elastic and to limit the possibility of queries against large data sets taking significant resources to process. For example, some cloud databases do not allow joins across tables, and some do not support a true database schema. These restrictions are a major challenge to moving between cloud database vendors, especially for applications built in a true relational model.

1.2.1.8 CCUC 3: Changing VM Hosts

This simple scenario addresses the case in which a cloud customer wants to take virtual machines, built in one cloud vendor's system, and run them in another cloud vendor's system. The main requirement of this operation is a common format for virtual machines.

1.2.1.9 CSC-CB: Cloud Bursting

The cloud bursting use case, focusing on interoperability issues at the IaaS level, describes a scenario where multiple cloud platforms need to work together. In particular, being it the typical situation, the use case describes the collaborations between public and private clouds and the possibility to move work and data loads between them. The use case illustrates a situation in which a private cloud, running one or more virtual machines, needs more computational power from a public cloud in order to respond to a peak of incoming request from customers or to speed up computation. In order to do so, VMs are dynamically migrated from one environment to another. The use case is similar to the CCUC 3: Changing VM Hosts scenario, but it differs from it not only for the dynamic requirements imposed, but also because the migrated VMs are supposed to collaborate with the one still hosted on the source cloud platform.

1.2.2 A Case Study

Let us consider a typical architecture of a business intelligence application in which an ETL process (extraction, transformation and loading) retrieves data from a database (DBMS), a customer relationship management (CRM) and an enterprise resource planning (ERP) system, and preprocesses them for further analysis after their storage in a data warehouse system (See Fig. 1.6).

The CRM and ERP components use data coming from their own databases. The data recorded in the data warehouse are used by the OLAP system for the creation of business reports on sales, marketing, management, budget, and so on. The data mining component uses the same data to perform market analysis to identify new product bundles to find the root cause of manufacturing problems, to prevent customer attrition and acquire new customers, to cross-sell to existing customers, and to profile customers with more accuracy. The first concern arising when we introduce such an architecture in the cloud computing environment is interoperability, since several application components need to interoperate to achieve the business goal. This could introduce problems when such components are distributed among several clouds: the cloud services hosting the components must share a compatible programming interface in order to avoid rewriting the entire application. Furthermore, nothing prevents cloud platform providers from limiting access to offered storage services to applications residing on their own platform. Another important aspect to handle

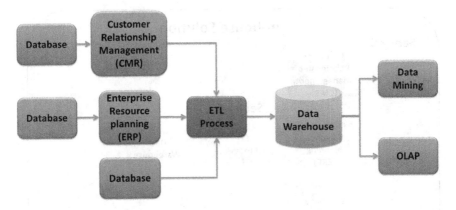

Fig. 1.6 Architecture of a Business Intelligence application

is migration to and portability among clouds. Suppose that the application is built in a particular cloud service or on an in-house system and, for cost, performance, or other reasons, the owner wishes to move it to another service provider. In a fully interoperable setting, the application could access data from both in-house and public cloud databases through a common interface. In a real situation, differences in API, data, and message formats or communication protocols represent a concrete obstacle to achieve such a feature.

1.2.2.1 Motivating Example: Portability and Interoperability Consideration

The most important dimension to consider is the Service Level. Starting from the lower value of this dimension, we can consider the case in which the application is built in an in-house IT and its owner wants to migrate it to the cloud (Fig. 1.7). We need to distinguish the target cloud service models because the problems that arise are different. In the case of a target IaaS platform, the requirements are to package the entire software stack in a set of VM images and load these images on the target resources acquired. The migration process in this case may be prevented by incompatibilities between virtual image formats or particular requirements at the infrastructural level. In our example, the software (ERP, CRM, ETL, data warehouse) delivered on-premise in the in-house solution will be packaged jointly to the entire software stack. Additional storage systems may be attached to the VMs to offer storage capabilities to the databases and Data Warehouse. In case of migration to a PaaS, the processes constituting the application can be deployed in the platform, provided that a compatible environment is offered. The databases, the Data Warehouse, the ERP and CRM module can be supplied by platform services if present in the platform or offered in a multicloud environment by other providers at the SaaS level. In

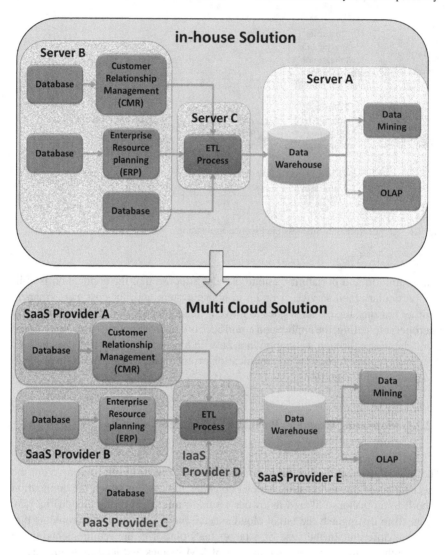

Fig. 1.7 Porting of an in-house system to multiple clouds

this case a certain importance is held by the compatibility of the functional interface between the old and new software.

When porting occurs among providers at the same level, the main concerns are related to the lack of common interfaces to access the different services. For instance, at the PaaS level cloud vendors generally provide their own solutions and APIs for handling databases. Consider that the databases of the ERP and CRM can reside on different platforms: if such platforms were fully interoperable, the ERP service would be able to freely access information regarding all the databases. But rarely, if ever, this happens because of the lack of common interfaces, for security reasons or simply due to commercial strategies adopted by cloud providers. Interface standard-ization is required for porting applications/software systems to the cloud, or among cloud platforms, in order to enable applications to use services and protocols pro-vided by the platform(s), as well as to provide access to the capabilities supporting the application. This means that, even when standard languages are used, portabil-ity at the PaaS level is not guaranteed, since implementations of platform services may vary between providers. Interoperability issues may arise even when portability is achieved. Assuming that portability among clouds is feasible and that the appli-cation is distributed among several clouds, there is still no guarantee that various parts of the application are able to interoperate, due to a lack of shared data format, communication interfaces, security requirements, or providers' commercial policies. Another interesting example is represented by the case of a **Hybrid Cloud** solution, in which some resources are offered by a **Public Cloud**, for example the databases and the ERP process, while other critical components, such as the data mining and OLAP processes, are internally managed by the organization following a **Private Cloud** model. This approach allows organizations to take advantage of the scalabil-ity and cost-effectiveness of public cloud offers, without exposing mission-critical applications and data, which can be dealt with in-house. However, no matter how similarly a public and private cloud are built, design and implementation differences will inevitably exist and interoperability issues may occur. For what concerns issues related to data portability which can arise in our example application, if the data are not stored in a PaaS storage but are stored in a database installed on a VM, the problem of incompatibility of the export formats of data is certainly minor. Consider instead the possibility that the ETL process uses data stored in a three columnar database offered by a cloud provider A, or even three different cloud providers A, B, C for redundancy. What if, for security reasons, the application needs to migrate data to a new provider X? What happens if the cloud providers A, B, C, and X do not support the same export and import formats? And even more, can we assume that provider A offers export mechanisms? The situation is even worse if the data are stored at the SaaS level, because in this case the format and the content of the cloud service customer data are totally in the hands of the cloud service provider. Suppose that our business application works for a very large enterprise and, for this reason, it manages a lot of data, needs to be decentralized and some of its components need to be replicated over different providers for security. Referring to our motivating exam-ple, we can have a duplicated data warehouse hosted by different providers and data mining and OLAP processes that would like to access dynamically the most available storage each time. This scenario surely requires the use of a common API to access the different data warehouse copies, but this situation involves also the compatibility of data representation (not to mention problems related to data localization).

References

1. Mell, P., Grance, T.: The NIST Definition of Cloud Computing Recommendations of the National Institute of Standards and Technology. Computer Security Division, NIST, Gaithersburg, MD (2011)
2. Behrendt, M., Glasner, B., Kopp, P., Dieckmann, R., Breiter, G., Pappe, S., Kreger, H., Arsanjani, A.: Introduction and architecture overview—IBM Cloud Computing Reference Architecture 2.0 (2011)
3. Buyya, R., Vecchiola, C., Selvi, S.T.: Mastering Cloud Computing—Foundations and Applications Programming. Morgan Kaufmann/Elsevier, Waltham, MA (2013)
4. The Opengroup Consortium: http://www.opengroup.org
5. Baudoin, C., Dekel, E., Edwards, M. et al.: Cloud Standards Customer Council. Interoperability and Portability for Cloud Computing: A Guide (2014). http://www.cloudstandardscustomercouncil.org/CSCC-Cloud-Interoperability-and-Portability.pdf
6. Ahronovitz, M., Amrhein, D., Anderson, P. et al.: Cloud Computing Use Cases-white paper. http://cloudusecases.org/Cloud_Computing_Use_Cases_Whitepaper-4_0.odt

Chapter 2
Methodologies for Cloud Portability and Interoperability

2.1 Model-Driven Approach for Design, Provisioning, Execution, or Migration to the Cloud

The OMG model-driven architecture (MDA) [1] is a model-based approach for the development of software systems that aims at separating the platform-independent design of a software application from its implementation on a given platform. The main feature and benefits of MDA from the cloud perspective are the enablement of portability, interoperability, and reusability of (parts of) the system, as well as its easy maintenance, through human-readable and reusable specifications at various levels of abstraction. In the context of cloud computing, model-driven development allows developers to design software systems in a cloud-agnostic way, and to be supported by model transformation techniques into the process of instantiating the system into specific and multiple clouds. This approach, which is commonly summarized as "model once, generate anywhere", is particularly relevant when it comes to designing and managing applications across multiple clouds, as well as migrating them from one cloud to another. Combining model-driven application engineering and the cloud computing domain is currently the focus of several research groups and projects, among others, MODAClouds [2], ARTIST [3], PaaSage [4], and REMICS [5], which we are briefly illustrating in the following.

2.1.1 MDA in MODAClouds

MODAClouds—MOdel-Driven Approach for design and execution of applications on multiple Clouds [6] is an EC-funded research project that proposes a model-driven approach aimed at supporting system developers and operators in utilizing multiple clouds and in migrating parts of their systems from cloud to cloud as needed. The MODACloudML platform relies on a domain-specific language for the design and execution of applications on multiple clouds. The model-driven engineering approach adopted by the MODACloudML platform allows developers

© The Author(s) 2015
B. Di Martino et al., *Cloud Portability and Interoperability*,
SpringerBriefs in Computer Science, DOI 10.1007/978-3-319-13701-8_2

to build the system at various levels of abstraction. The three envisioned levels are: the *Cloud-enabled Computation Independent Model* (CCIM) to describe an application and its data, the *Cloud-Provider Independent Model* (CPIM) to describe cloud aspects related to the application in a cloud-agnostic way, and the *Cloud-Provider Specific Model* (CPSM) to describe the cloud details needed to deploy and provision the application in a specific cloud. Each layer of the architecture contains various models that can be manipulated within the MODACloudML environment.

MODAClouds provides developers and operators with the following tools to support the application lifecycle management:

- the decision-making toolkit to compare and analyze different cloud solutions;
- the IDE to support a cloud-agnostic design of software systems, a partial code generation, and the deployment in the selected target;
- the runtime layer to monitor the execution of the system on multiple clouds.

2.1.1.1 Engagement with Case Study and Positioning with Respect to Use Case Scenarios and Features

To evaluate the real usability of the MDA approaches, we try to use the MODAClouds on the application example reported in Fig. 1.6. Following the MODAClouds methodology, each part of the application can be represented by a generic resource with specific needs and capabilities (Fig. 2.1).

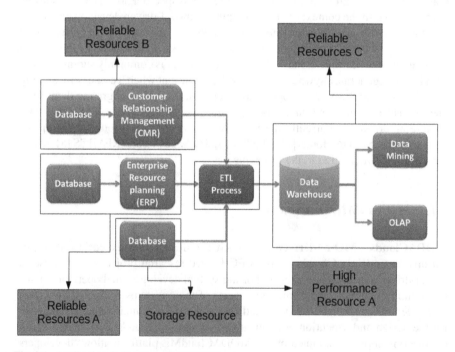

Fig. 2.1 Application design through MODAClouds

Every resource in the model created above can be replaced by a service offered by different cloud providers. Through the application of the concept "Model Once, Generate Anywhere", MDA enables the portability from a cloud provider to another, dramatically reducing the complexity of developing an application all over again. One of the key features of these methodologies is the support of high-level design, early prototyping, and semiautomatic code generation. Considered as the high level of abstraction, the porting of an application modeled using MDA to a cloud provider cannot be done without the use of existing or emerging standards: such as TOSCA. These kinds of standards are essential for transforming a platform-independent model to a platform-specific model. This approach can be helpful in the scenarios CSCC S1 and CSCC S2. In particular, the MODAClouds IDE will support a cloud-agnostic design of software systems, the semiautomatic translation of design artifacts into code, and their deployment in the selected target clouds. Some support to interoperability is provided by some components such as the multicloud load balancer and the resource allocator. The first is a load balancer used by the applications and handles the case when the same application runs in two clouds (as distinct instances), and which is able to route traffic toward the multiple instances. The resource allocator is a centralized service, mediating resource allocations from various cloud providers, enabling asynchronous interactions, and tracking. MODAClouds-targeted applications are mainly the PaaS and IaaS. The positioning of this solution according to our dimensional space is illustrated in Fig. 2.2.

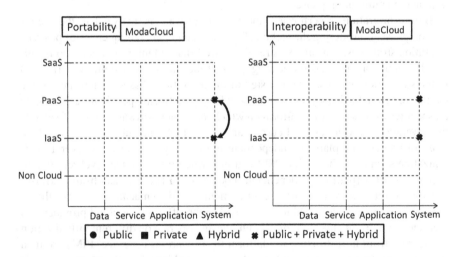

Fig. 2.2 MODAClouds solution positioning

2.1.2 MDA in ARTIST

ARTIST—Advanced software-based seRvice provisioning and migraTIon of legacy SofTware [7] is an EC-funded research project that proposes a software modernization approach covering business and technical aspects. In particular, ARTIST employs model-driven engineering techniques to automate the reverse engineering of legacy software and the forward engineering of cloud-based software in a way that modernized software truly benefits from targeted cloud environments.

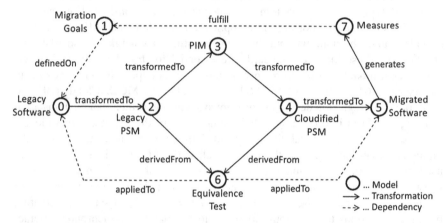

Fig. 2.3 The ARTIST software modernization process [8]

The proposed modernization process (Fig. 2.3) consists of a premigration, migration, and postmigration phase.

Before migration is performed, the legacy software is analyzed in the premigration phase, considering the technical and nontechnical consequences of possible migration strategies. This analysis results in well-defined migration goals constituting the input for the decision-making on how the migration should be performed in the subsequent phases. In a first step of the migration phase, models are reverse engineered from the legacy software. These legacy platform-specific models include all specifics imposed by the platform on which the legacy software is built. To enable the coverage of a wide range of current and future modernization scenarios and the reuse of reoccurring platform-independent migration patterns across several modernization scenarios, the legacy PSM is transformed into a higher level representation, called PIM (platform-independent model). The PIM abstracts from platform-specifics, such as software runtime environments and data management capabilities. These platform-specifics need to be adapted to the offerings of cloud providers, as their cloud environments are typically unique and operate on different virtualization layers, i.e., from infrastructure to platform to software as a service. PIMs are then subject to model transformations, which are selected based on the migration goals defined in the premigration phase.

These transformations implement the actual migration by applying optimization patterns and integrating cloud-specific modernization opportunities. As a result, model-based representations of the migrated software that include platform-specifics compatible with the selected cloud environment are produced. Such a cloudified PSM is transformed into the executable migrated software hosted in a cloud environment.

In the postmigration phase, model-based representations of the legacy and the migrated software are employed to derive equivalence tests. They aim at verifying that the migrated software behaves as expected. Furthermore, nonfunctional properties are evaluated to certify if the migration goals are fulfilled. This is achieved by analyzing the execution of the migrated software to obtain quality measures that are checked against the defined migration goals.

2.1.2.1 Positioning with Respect to Use Case Scenarios and Features

ARTIST accomplishes the use case scenario CSCC S5. In fact, this solution has been implemented to enable the effective migration of legacy applications to cloud environments. The ARTIST methodology comprises all the phases needed to perform modernization. It defines the life cycle of the migration process including its phases, purpose, activities to be performed, inputs, outputs, tools, or techniques suited for each phase and the templates to use. Several software tools are provided by the project to support the methodology and assist the developers in the migration process. Reverse engineering tools, cloud metamodeling language, forward engineering tools, testing framework, and technical/business feasibility tools are some of the tools provided by the project. The positioning in our dimensional model of this methodology is illustrated in Fig. 2.4.

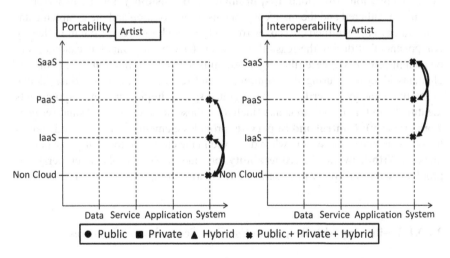

Fig. 2.4 ARTIST solution positioning

2.1.3 MDA in REMICS

REMICS—Reuse and Migration of legacy applications to Interoperable Cloud Services [9] is an EC-funded research project aimed at supporting the migration of legacy systems to service clouds by providing a model-driven methodology and tools.

The software products are built with subsequent model refinements and transformations from business models (process, rules, motivation), down to component architectures (e.g., SOA), detailed platform-specific design, and finally implementation. REMICS proposes to improve existing approaches and extend them when needed, to provide a holistic view of migration that covers the whole process with a methodology, tools, languages, and transformations. One main objective of REMICS is to provide its solutions based on standards and open-source tools as much as possible to facilitate reuse and shorten the barrier for users to take advantage of the innovations.

The project intends to significantly enhance this generic process by proposing a set of advanced technologies for architecture recovery and migration, including innovative technologies such as model-driven interoperability and Models@Runtime.

Model-driven interoperability is a rather new domain, which builds on top of a long history on data and service interoperability. Semiautomated methods that assist users to handle interoperability issues between services are also addressed in REMICS. In general, the REMICS migration methodology is focused mainly on the evolution of the technology model. There are seven activity areas defined in the REMICS methodology, which cover the full life cycle of a legacy system modernization and migration to the cloud: **Requirements and Feasibility** (requirements for the system are gathered and the main components of the solution and their implementation strategy are identified); **Recover** (recovery of the knowledge from those legacy components that during the feasibility analysis have been pointed as candidates to be reengineered); **Migrate** (the target system is defined and implemented using the elements identified during the requirement and recover phases); **Validate** (define the testing strategy to verify that the migrated system implements the requirements identified and that the components (including those not reengineered) and services work properly); **Control and Supervise** (provides elements to monitor and control the performance of the system when deployed in the cloud and to modify that performance); **Withdraw and Interoperability** (provides tools that solve interoperability problems with third-party providers or any external components and services).

2.1.3.1 Positioning with Respect to Use Case Scenarios and Features

The REMICS project falls under the objectives described in the use case scenario CSCC S5, but covers also some objectives of use cases CSCC S4 and CSCC S2.

The migration process defined in REMICS consists of understanding the legacy system in terms of its requirements, architecture, and functions, designing a new SOA that provides the same or better functionality and quality of service, and verifying and implementing the new application in a cloud computing platform suitable for the purpose. The positioning in our dimensional model of this methodology is illustrated in Fig. 2.5.

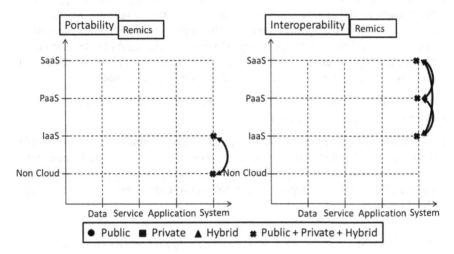

Fig. 2.5 REMICS solution positioning

2.1.4 MDA in PaaSage

PaaSage—Model-based Cloud Platform Upperware [4] is an EC-funded research project that aims at delivering a development and deployment platform, sustained by a proper methodology, through which software engineers, and in particular developers of enterprise systems, can access services of cloud platforms in a technology neutral environment. The platform will abstract all the technical details, while guiding the developer during the entire life cycle of her applications, allowing for a model-based development and an optimal configuration of the software, independent from of the underlying cloud infrastructure. This will enhance cloud portability since the application development, deployment, and execution will be independent of the reference cloud platform, and migration of obsolete or legacy applications to the cloud will also be supported.

2.2 Semantic Approaches

One of the contributory factors of interoperability and portability problems is the difference in the semantics of the resources offered, since no uniform representation exists. As stated in [10], semantic models are helpful in three aspects of cloud computing:

- Functional and nonfunctional definitions, that is, the ability to define application functionalities and quality-of-service details in a platform-agnostic manner;
- Data modeling, including meta-data added through annotations pointing to generic operational models, which plays a key role in consolidating API's descriptions;
- Service description enhancement, in particular regarding service interfaces that differ between vendors even if the operations' semantics are similar.

Existing technologies inherited from the Semantic Web field can be useful to address these aspects. In particular:

- **Web Ontology Language (OWL)** [11] can define a common, machine-readable dictionary that is able to express resources, services, APIs and related parameters, service level agreements, requirements, offers, and related key performance indicators (KPIs). Listing 2.1 reports an example of OWL description for a "Compute" cloud resource.

```
<http://publicaddress.org/ontology#CloudResource>
rdf:type owl:Class ;
rdfs:subClassOf <http://publicaddress.org/ontology#
    Resources> .
<http://publicaddress.org/ontology#Compute> rdf:type
    owl:Class ;
rdfs:subClassOf <http://publicaddress.org/ontology#
    CloudResource> .
<http://publicaddress.org/ontology#cpu> rdf:type
    owl:DatatypeProperty ;
rdfs:domain <http://publicaddress.org/ontology#
    Compute> ;
rdfs:range xsd:integer ;
rdfs:subPropertyOf owl:topDataProperty .
```

Listing 2.1 OWL description of a cloud resource

- **OWL for Services (OWL-S)** [12] adds semantic to cloud services in order to enable users and software agents to automatically discover, invoke, and compose them.
- **SPARQL** [13] is an RDF query language that is able to retrieve and manipulate data stored in resource description framework format on which OWL is based. SPARQL queries can be used to retrieve RDF described resources, filtered according to selected constraints. Listing 2.2 reports an example of a SPARQL query, executed on the same database used for example in Listing 2.1, where only computing resources with a minimum number of CPUs are selected.

```
SELECT ?Compute_Resource   ?CPU_Number
WHERE {?Compute_Resource rdf:type ontology:Compute.
       ?Compute_Resources ontology:cpu ?CPU_Number.
       FILTER (?CPU_Number >=3)}
```

Listing 2.2 SPARQL query executed on an RDF database

- **Semantic Web Rule Language (SWRL)** [14] expresses additional rules and heuristics.

Owing to the aforementioned potential of semantic technologies, a number of works represent cloud resources, services, and in general cloud concepts in OWL producing so-called *Cloud Ontologies*. Particularly in [15] an ontology is proposed which focuses on the technologies involved in the cloud phenomenon and describes the different layers of cloud computing, the relationships between them, and the users of each cloud layer, while in [16] an ontology is proposed built on existing standards, developed to improve interoperability between existing cloud solutions, platforms, and services, both from end user and developer sides. In [17] a unified OWL of cloud resources is described at the PaaS and SaaS levels, which focuses on the classification and categorization based on a functional analysis, of cloud services and virtual appliances. In [18, 19], the description of functional and nonfunctional characteristics of some specific cloud services is proposed, alongside information related to exchanged parameters and collaboration between services.

2.2.1 Semantics in mOSAIC

The EC (FP7)-funded project **mOSAIC—Open-Source API and Platform for Multiple Clouds** [20] addresses the issues related to cloud portability and interoperability with a number of technologies. In particular, it applies semantic technology in two components of the mOSAIC framework, the semantic engine [21] and the dynamic discovery and mapping system [22].

The **Semantic Engine** and associated ontologies were developed to support the cloud application developer in the tasks of discovering the needed functionalities and resources for application development through vendor-independent representations of such application components, and representation of generic programming concepts and patterns, including application domain related ones. The semantic engine introduces a high level of abstraction over cloud APIs and cloud resources, by providing a semantic-based (namely OWL) representation of abstract functionalities and resources, related by properties and constraints, and application domain level concepts and application patterns. Inference rules representing developer experts' knowledge and reasoners are also used.

The semantic engine overcomes syntactical differences representing the API semantically, independently from the programming model. It offers a catalog of functionalities related to the cloud domain, and patterns related to design and specific application domains, representing services and resources in an agnostic way. From the viewpoint of the developer, the application is designed (by using the associated GUI) starting from the application domain concepts (which are not related to cloud computing).

In such a way, the developer can reach a suitable design for the cloud application. Following this, the developer can use the semantic engine to obtain the actual application descriptors needed by the mOSAIC platform to successfully deploy the application in a selected IaaS provider.

The **Dynamic Discovery Service** is the mOSAIC answer to the need for automating discovery mechanisms and alignment facilities due to the growing number of cloud providers that deploy their offers. A possible solution to this problem is proposed through the application of semantic and matchmaking technologies.

The Discovery and Mapping Service's target is to discover cloud providers' functionalities and resources, compare and align them to the mOSAIC API, thus supporting agnostic and interoperable access to cloud providers' offers. This module of the mOSAIC framework is mainly based on and supports already existing languages for the semantic description of Web services, and it utilizes both node level and structural level matching to discover the mapping between cloud providers' services and mOSAIC APIs.

2.2.1.1 Engagement with Case Study and Positioning with Respect to Use Case Scenarios and Features

The semantic engine's main aim is to support the user in selecting cloud APIs' components and functionalities needed for building the application in the cloud, and the list of needed resources to be acquired from the cloud providers. Thus, we can apply our case study (Fig. 1.6 in the context of the use case scenarios illustrated in Sects. 1.2.1.5 and 1.2.1.1). The application example contains a structure that invokes a pipe and filter style. Through the use of the semantic engine, it is possible to describe semantically the application by using an application pattern (Fig. 2.6). This application pattern can be mapped on specific mOSAIC components and API or on other providers equivalent to API. In a scenario in which the application is built in an in-house system, the semantic engine suggests the component to use to deploy the application in the cloud, while if the application is already in the cloud the engine will suggest, by using inference rules and the implemented discovery mechanisms, the components of other providers to be useful to replace the in-use components. The positioning of this solution in our n-dimensional space is illustrated in Fig. 2.7.

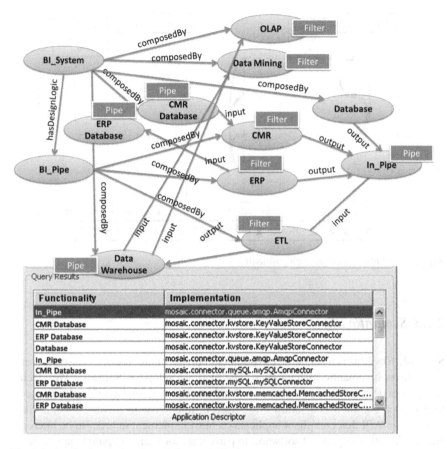

Fig. 2.6 Application design through Semantic Engine

2.2.2 Semantics in Cloud4SOA

The EC-funded project **Cloud4SOA** [23] focuses on resolving the semantic interoperability issues that exist in current cloud infrastructures and on introducing a
user-centric approach for applications which are built upon and deployed using cloud
resources. The aim of Cloud4SOA is to facilitate cloud-based application developers
in searching for, deploying and governing their business applications on the PaaS
offerings that best match their needs. Additionally, through the semantic interconnec-
tion of heterogeneous PaaS offers, the framework supports the switching between
PaaS providers. Cloud4SOA supports searching between the existing PaaS offerings for those that best match the developers' needs. The matchmaking algorithm
computes the degree of similarity between the semantic descriptions of the PaaS
offerings and application profiles. The Cloud4SOA framework aims to support a
seamless migration between platforms, tackling semantic interoperability conflicts
to allow for portability of an application and its data.

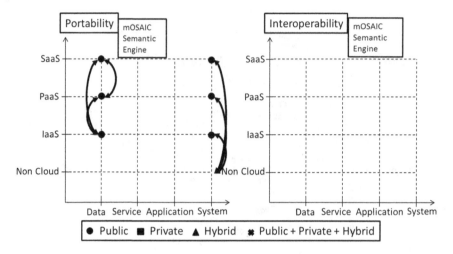

Fig. 2.7 Semantic Engine solution positioning

2.2.3 Semantic Sky

Semantic Sky [24] is a semantic environment born to interconnect cloud-based services through semantic technologies. The platform relies on Semantic Web technologies to deliver a uniform vision of cloud services and resources, enabling users to exploit a common and familiar interface to interact with them. Semantic Sky provides an API that compliant applications must implement in order to be interoperable with other supported software. In particular, for each compatible application Semantic Sky provides an **Application Plug-in**, used to determine the user's working context and infer additional meta-data, when possible. In order to do so, the platform architecture includes a set of components which analyze the data provided by the plug-ins, generally in the form of a text string, annotate them, and infer new knowledge.

- The **Knowledgebase** represents an OWL/RDF database where all user data are stored and indexed. SPARQL queries can be executed on it in order to retrieve meaningful information.
- The **Context Extractor Module** is the core engine of the system. It analyzes each token (word) the input data has been serialized into by a plug-in, and then determines all the resources connected to it, exploiting the index present in the Knowledgebase.
- The **Semantic Web Services Repository** contains semantically annotated SOAP and RESTful Web services, which are used to execute actions suggested to the user on the basis of the detected working context.
- The **Action Search Module** uses the list of resources detected by the Context Extractor Module as keys to research Web services inside the Semantic Web

Services Repository. If the previously detected resources only partially satisfy the input requests of some Web service, the system iteratively scans the directory in order to find other services offering the needed resources as output. Once all the inputs and outputs have been determined, the system proposes a set of Web services or **actions** to the customer to use.

- The **UI Generator Module** is responsible for the production of the graphical interface of the system, which must be suitable to display different document typologies.
- The **Module for Action Execution** is responsible for the correct execution of the semantically annotated actions.

With such an architecture, the system is able to integrate with existing, widely adopted applications and software, such as Gmail and Microsoft Office. One major benefit derives from the possibility to leverage a common, shared interface to interact with all the supported applications, reducing the users' learning time.

2.3 Multi-Agent Systems

Multi-Agent Systems (MASs) seem to offer one of the most effective approaches to solve a number of interoperability issues and automate a number of activities, in particular brokering, negotiation, management, monitoring, and reconfiguration in multiple clouds. MAS can be defined as a computerized system composed of interacting intelligent agents, collaborating within the same environment. An agent is an autonomous entity, represented by a software program, a robot, or even a human being. Despite the differences existing between cloud computing and multi-agent systems, their integration could provide solutions to problems arising in both environments. According to [25], agent-based solutions can improve cloud resources and service management and discovery, SLA negotiation, and service composition.

- As regards **brokering**, agents can access service and utility markets on behalf of users, retrieving the most updated resource configuration that satisfies applications' requirements. According to the available resources, SLAs are generated and services are booked from one or even multiple cloud providers. Their orchestration can be left to the user or executed by agents themselves.
- In case the users' constraints are too strict and no resource configuration can fully satisfy the requirements, agents can engage in several brokering rounds, loosening some constraints in order to deliver the best fitting SLA. In this situation, we talk about **negotiation** and agents with their dynamic adaptation ability, which are the perfect candidates for the job.
- **Monitoring** activities, focused on resource utilization and applications' requirements, can be performed by agents which can also renegotiate SLAs and/or optimize configurations (resources' **reconfiguration**) according to the changing requirements.

- When storage services are involved, agents can be used to search, filter, update, and query large volumes of data without the user's direct intervention.

2.3.1 Brokering, Negotiation, and Monitoring with mOSAIC's Cloud Agency

The mOSAIC project demonstrates, with its component **Cloud Agency** [26], the benefits of adopting such kind of a cloud multi-agent architecture. The mOSAIC's Cloud Agency is a service for the deployment and execution of mOSAIC application. It is in charge of provisioning, from different providers, a collection of cloud resources, which fulfills at best the user's requirements, to be consumed by mOSAIC applications. The cloud agency aims at advancing the state of the art of using the clouds by providing a decision-making support to the user for discovery and decision about the best cloud solution that satisfies his requirements. The cloud customers need to detect underutilization and overload conditions, and to make decisions on load balancing and resource reconfiguration. The cloud agency aims at providing a monitoring service that runs on IaaS under the control of the customer. Implemented as a multi-agent system [27], the cloud agency is based on asynchronous messaging, as other mOSAIC software prototypes. It offers a RESTful interface compliant with OCCI [28]. The brokering of the best collection of cloud resources has been modeled as a multi-criteria optimization problem, with hard and soft constraints that can be included by the user in the Call-for-Proposals [29]. Beyond the provisioning role, the cloud agency also has other resource management functionalities, like monitoring, which is related to the parameters specified in SLAs (monitoring the quality of service). Details about this role implementation can be found in [11].

2.3.1.1 Engagement with Case Study and Positioning with Respect to Use Case Scenarios and Features

Multi-agent systems represent a meaningful support in multi-cloud environment due to their ability to automatize operation such as brokering, negotiation, monitoring, and reconfiguration. In particular, suppose that we have developed the application example of Fig. 1.6 on a particular cloud or multi-cloud setting. Suppose that this solution is implemented and hosted in an IaaS solution of a certain provider A. Using the cloud agency [11] it is possible to monitor the resources in which the application is deployed through mobile software agents that take measures inside the cloud resources, make decisions based on monitoring values or using automatic settings, and eventually decide to migrate the application to another IaaS provider. An example and the positioning of this mOSAIC component is illustrated in-depth in Sect. 3.5.1.

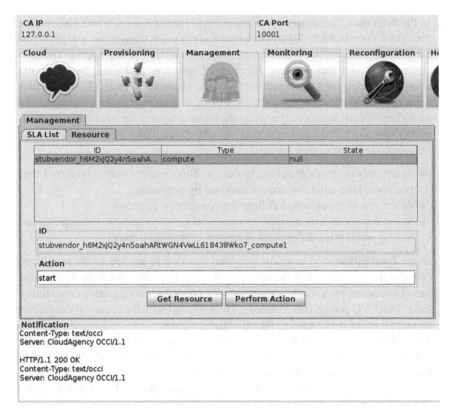

Fig. 2.8   Cloud agency management GUI

During the migration phase it is possible to take advantage of the multi-agent system to manage with agnostic interface resources of different providers (Fig. 2.8 through vendor agents which implement wrappers for specific clouds), negotiate with other providers, and reconfigure the application in the new chosen cloud provider.

2.3.2 Agent-Based Cloud Resource Management Testbed

The work presented in [30] shows an interesting example of how multi-agent systems can be used to manage resources in a cloud computing environment. In the proposed testbed, both cloud providers and consumers are represented by agents acting as their intermediaries. A discovery process, composed of four steps called *selection, evaluation, filtering, and recommendation*, brings broker agents to match requests from users with the providers' offers.

1. In the **Selection**, consumers submit requests to broker agents, which also receive providers' specifications about the resources they offer. Consumers and providers do not communicate with brokers directly, but through proxy agents. The consumers' requests may contain different information, including the maximum acceptable price for a resource, its type and characteristics (CPU power, disk capacity, bandwidth), the time slot for its use, and so on. On the other hand, the providers' specifications may contain the minimum price for a resource, its availability time slots, and a description of its capabilities. Three matching criteria can be applied to the selection process:

   - Select only perfectly matching resources. This criterion may not return results because of the strict constraints imposed by consumers.
   - Select resources closely fulfilling the consumers' requests.
   - Select resources with availability time slots corresponding to the requests and having small price differences in respect to the requirements.

2. In the **Evaluation** step, a utility weighted function is applied to the resources' configurations retrieved in the selection stage in order to present to the user the most convenient ones. Each characteristic of a resource configuration (availability time slots, CPU power, disk volume...) is compared with the requests' details and an overall score is assigned to the potential matching.

3. The **Filtering** stage consists of applying a selection criterion to the potential matches, to which the utility score has been previously assigned. In particular, the system automatically refuses matches with the lowest score and, among the remaining results it discards the ones below a certain threshold previously set by the consumer.

4. Since the testbed is based on a multi-broker agent system, with each broker having access to different providers' specifications, it is possible that one or more of them cannot find a useful resource configuration to fulfill consumers' requests. In this case, a broker can send a **Recommendation** to the consumer agent, informing that other brokers may answer the request.

2.4 Cloud Patterns

According to a definition provided by IBM in [31], cloud computing patterns are "logical descriptions of the physical and virtual assets that comprise a cloud computing solution". They have arisen from the need to provide both general and specific solutions to recurring problems in the definition of architectures for cloud applications: indeed, while classical design patterns deal with problems related to different aspects of software development, being the structural view of the application only one of these aspects, cloud patterns, mainly focuses on the architecture of the cloud solution. This has led, in many cases, to the development of platform-dependent patterns, which can be applied only to a specific platform offered by a specific vendor. Various existing cloud patterns' catalogs, given their different nature and objectives, deliver content at different levels of detail and abstraction. Some patterns, like those presented in [32, 33], are tied to a specific cloud platform, thus being more detailed in terms of the components they rely on for the implementation, but the solution they provide is strictly dependent on the reference platform and has poor flexibility. Catalogs developed by academic efforts, like those defined in [34, 35], are not tied to industrial proprietary solutions: they describe general functionalities and behaviors and they propose architectural models that are much less bound to specific cloud platforms, thus resulting in less details and better flexibility. For this reason we refer to them using the general term of "Agnostic Patterns". Here, we provide a brief introduction to the main cloud pattern-based approaches and cloud patterns' catalogs, with particular emphasis on those published by Amazon for AWS and on the agnostic patterns.

2.4.1 How Cloud Patterns Can Enable Interoperability and Portability

Despite the poor flexibility shown by some vendor-specific patterns, cloud patterns still represent a valuable means to enhance portability and interoperability between cloud platforms. First of all, patterns can be used to describe and model existing cloud applications in a very easily understandable manner, tracing back the different cloud implementations to a set of well-known and stable solutions. In this way, it becomes easier to understand the exact functionalities and responsibilities of a specific cloud application component, which can be at a later time substituted with a compliant one having the same or similar characteristics. This approach can be also exploited in the case of porting of non-cloud application, describable through classic design patterns, to a cloud environment, provided a mapping between design and cloud patterns' participants exists. Furthermore, using a cloud pattern, and in particular an agnostic one, as a canvas on which to develop a new application, it would be possible to implement each of the pattern's participants with services and components exposed by different cloud vendors.

2.4.2 IBM Virtual Patterns

According to IBM [36], a pattern is a collection of elements describing a complete and fully functional software solution, which can involve different interconnected systems or a single entity. All the knowledge needed to create, configure, and support every aspect of the solution is already included in the pattern. As a result, IBM has developed four categories of patterns [37], whose definitions are reported here unchanged:

- **Virtual System Patterns** (VSPs) are repeatable topology definitions that are based on various virtual images, each containing multiple middleware components and applications that are configured to work with each other.
- **Virtual Application Patterns** (VAPs) offer a view of a virtual application with which the user can focus only on application requirements and not on the underlying infrastructure that is needed to support the runtime environment.
  A VAP represents a collection of applicative components, behavioral politics, and relative links. A workload service, which takes advantage of this kind of pattern, can automatically build the infrastructure and middleware resources needed to operate and manage the virtual application, which thus becomes an instance of the pattern.
- **Database Patterns** are IBM DB2 product extensions that are used to build DB2 databases linked to a virtual application as an existing database component. The existing database component can be a database pattern instance that is managed within the cloud environment, or it can be a remote database that was created and is managed outside the cloud environment.
- **Virtual Appliance Patterns** are VMs that consist of a single-server workload instance with a preconfigured operating system and all of the middleware, applications, and script packages necessary to automatically deploy and configure the application environment.

In order to use a virtual application pattern, a customer can exploit the pattern builder service offered by a platform, provided by IBM itself and known as **Workload Deployer**, which can be accessed through both the IBM **PureApplication System** (Private Cloud) and the **Smart Cloud Application Workload Service** (Public Cloud), based on the IaaS platform provided by the **IBM Smart Cloud Enterprise**. Figure 2.9 shows the tool used to manage a virtual application platform in the IBM Smart Cloud Application Workload service. The reported VAP is used to build Web applications and it is composed of three elements:

- An **enterprise application** component, where the actual application resides. By clicking on the component a user can add the artifacts needed to run and manage the application.
- A database component, representing the storage service associated to the application. A client can easily customize the database characteristics by accessing the component's properties, or she can leverage a database pattern to design it.
- A user registry element represents an existing LDAP service that can be attached to a Web Application component or to an Enterprise Application component to manage user accesses and privileges.

All the links existing between couples of components can be customized in order to specify how they can collaborate or access each other. Through the services provided by IBM, a user is free to use an existing virtual pattern as it is, just by providing the artifacts, which are necessary to configure and run applications and databases, or

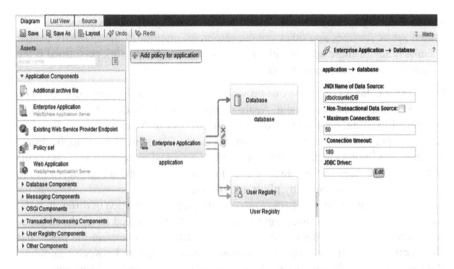

Fig. 2.9 Example of IBM VAP: enterprise Web application pattern

she can modify an existing pattern by adding and removing components. Available components can be chosen from a palette (shown on the left side in Fig. 2.9): the range of selectable elements varies according to the pattern type being edited. Pattern components can also be created from scratch using the Plug-in Development Kit (PDK) delivered by IBM itself.

2.4.3 Azure Cloud Patterns

The pattern catalog published by Windows Azure [33] represents a remarkable contribution to the development of cloud applications and components driven by patterns. The catalog currently contains 24 design patterns and 10 related guidance topics, accurately describing the benefit of applying patterns to application design, showing how each participant is mapped to a specific platform component. For almost all Patterns, code samples and snippets are provided and the benefits are discussed in-depth. The catalog currently comprehends 24 patterns. Some examples of these patterns are mentioned below:

- **External Configuration Store Pattern** Move configuration information out of the application deployment package to a centralized location.
- **Health Endpoint Monitoring Pattern** Implement functional checks within an application that external tools can access through exposed endpoints at regular intervals.
- **Pipes and Filters Pattern** Decompose a task that performs complex processing into a series of discrete elements that can be reused.
- **Priority Queue Pattern** Prioritize requests sent to services, so that requests with a higher priority are received and processed more quickly than those of a lower priority.

- **Scheduler Agent Supervisor Pattern** Coordinate a set of actions across a distributed set of services and other remote resources, attempt to transparently handle faults if any of these actions fail, or undo the effects of the work performed if the system cannot recover from a fault.

While other catalogs (e.g., the one provided by Amazon) define new patterns to be applied in the cloud environment, Azure's solutions are to be considered as specializations of existing, well-known, design patterns, to the Azure platform. This is a notable feature, since it is easier to map elements of existing, legacy applications, designed and implemented according to well-known design patterns, to cloud components and services. An example of the classical design pattern considered in the Azure catalog is represented by **pipes** and **filters**.

The main objective of this pattern is to modularize the execution of different tasks on (a stream of) data, so that single modules can be rearranged or modified separately without affecting the others. Two main participants compose the pipes and filters pattern: the **pipe**, constituting the communication means between computing elements and generally represented by some kind of Queue, and the **filter**, representing the computation unit that executes the required task(s). An implementation can instantiate any number of pipes and filters, each pipe connecting two filters and each filter dedicated to the execution of one or more tasks. The policy adopted by pipes to pass data from one filter to another strongly depends on the implementation. The example reported in Fig. 2.10 shows the solution proposed by Azure. Such a solution maps filter instances with an Azure "WorkerRole", while pipes can be implemented through a "Service Bus". By exploiting the similarities in features exposed by components provided by different cloud platforms, it would be easy to instantiate the pipes and filters pattern using heterogeneous services, since a clear description of its elements has been already provided in a "cloudified" form.

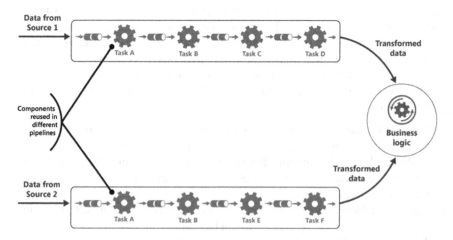

Fig. 2.10 The pipes and filters pattern in Azure

2.4.4 Amazon Web Services (AWS) Cloud Design Patterns

The AWS Cloud Design Pattern [32] catalog is a collection of solutions and design ideas that use the AWS cloud technologies to solve common design problems. These solutions are strongly connected to the AWS platform, so they are particularly well detailed and supported by precise diagrams that describe structure and interactions of the pattern efficiently. For each pattern there is a description that indicates the problems leading to the creation of the pattern and what difficulties can be resolved through its implementation. The implementation itself is described precisely, defining step-by-step the procedure of application of the pattern within AWS, specifying which components should be used and how this should be done. The whole catalog is organized into nine categories:

- **Basic Patterns** include basic solutions that can be completely reused or are partially involved in the definition of more complex patterns;
- **Patterns for High Availability** describe architectural solutions that can be used to provide highly available resources (including servers, data centers, or networks);
- **Patterns for Processing Dynamic Content** define how dynamic contents should be managed and processed in a distributed cloud architecture, using cloning and replication techniques, together with proxies and caches;
- **Patterns for Processing Static Content**, on the other hand, exemplify how static contents should be processed in a cloud environment, focusing on its distribution;
- **Patterns for Uploading Data** describe solutions for content upload in a cloud environment;
- **Patterns for Relational Database** provide solutions for the management of relational databases, with a focus on data replication and read-write optimizations;
- **Patterns for Batch Processing** focus on the execution of distributed, loosely coupled processes, managed through scalable queues;
- **Patterns for Operation and Maintenance** involve daily maintenance operations that should be performed without interrupting any running services and thus need swapping, replication, and monitoring capabilities;
- **Patterns for Network** describe solutions related to network maintenance and management, including load balance and firewall setup.

As an example extracted from the AWS catalog, let us consider the "Scale-Out Pattern": such pattern, included in the basic patterns' category, deals with issues related to resource scaling in case of processes working on high volumes of data or generating high volumes of traffic on a monitored network. The proposed pattern contains a brief description of the issues it deals with and a description of the proposed solution, together with a presentation of the platform components used to implement such a solution, in this case represented by:

- a load balancing service (**Elastic Load Balancing** or ELB);
- a monitoring tool (**CloudWatch**);
- an automatic scale-out service (**Auto-Scaling**);
- a set of Web services used to deliver compute capabilities (**EC2**).

A description of how these components have to be configured in order to interoperate is provided through both a textual and graphical description.

2.4.5 Agnostic Patterns: The CloudPatterns.org Community

`CloudPatterns.org` [34] is a community dedicated to the documentation of a catalog of patterns that includes a set of design solutions for the modularization of the relevant technological solutions for the configuration of cloud architectures. Patterns are here organized according to a hierarchical structure: the various descriptions are therefore not organized according to the purpose or functional area related to the problem addressed by the pattern, but essentially on the basis of a hierarchical composition of mechanisms and subpatterns, which are in turn defined. Despite the lack of a strong supporting graphical formalism, the different elements of each pattern and their iterations are described with good precision. A part of the catalog consists of "**Compound Patterns**", which take into account different models of provision and distribution of services (public and private clouds, IaaS, PaaS, and SaaS service models, and so on): such models are decomposed into sets of coexisting patterns, which define the key and optional features provided. Because a pattern often refers to others to describe the solution to a problem (the term *pattern language* is used to describe such a situation), each of them could be defined as *compound*. The patterns are applied through the implementation of technological *mechanisms*, which can be also composed of a set of interrelated mechanisms. Mechanisms are well-defined technological artifacts that have been established in the IT industry and generally refer to a certain computational model or platform. The nature of cloud requires the formal definition of a series of mechanisms that can be combined and implemented in different ways. It also requires the standardization of practices and solutions in the form of design patterns exploiting these mechanisms. Each provided cloud pattern is associated with one or more mechanisms. The catalog defines, in a specific section, the template used to describe each cloud pattern. However, compound patterns and mechanisms are not described according to a given, machine-readable, formalism, but are defined by a textual description sometimes supported by explanatory diagrams.

The components of the description are:

- **Requirement**: a simple and concise sentence that presents the basic requirements addressed by the pattern, in the form of a question.
- **Problem**: the causes of the problem addressed by the pattern and its effects are described in this section, which may be accompanied by a picture that illustrates it. Part of the description includes common conditions that lead to the birth of the problem.
- **Solution**: the design solution presented by the pattern to solve the problem and meet the requirements. Often the solution is only a short phrase that can be followed by a diagram that communicates the final solution in a concise manner. The details are provided in the "Application" section.

- **Application**: this part is devoted to describing how the pattern can be applied and may contain guidelines, implementation details, and sometimes the description of the entire implementation process.
- **Mechanisms**: this item contains a complete list of mechanisms that can be implemented to apply the pattern. Generally, some of the mechanisms will have been nominated in the "Application" section. The application of the pattern is not necessarily limited to the use of these mechanisms.
- **NIST Reference Architecture Mapping**: this section is provided for those who are interested in the way in which a pattern is related to the components that compose the National Institute of Standards (NIST) Cloud Computing Reference Architecture [38].

The representations of patterns describe different aspects of the pattern, such as its structure (components and relationships) and behavior (interactions between participants), explaining such descriptions often through simple diagrams and pictures, which use predefined icons. In general, the pattern does not describe exactly how individual components or mechanisms should be implemented with respect to a cloud platform: indeed many, if not almost all of the mechanisms can be implemented without any problems using an IaaS, PaaS, or SaaS model. This is also reflected in the structure of the compound patterns, which share many components.

2.4.6 Comparison Between Cloud Patterns

The vision of patterns of various providers is therefore different: providers like IBM see patterns as templates of applications which can be personalized through policies and constraints, on the basis of existing preconfigured solutions; Amazon offers a catalog of patterns that are expressed in terms of proprietary services which can be difficult to generalize to other cloud platforms; Azure, instead, applies existing design patterns to its cloud offering but, in general, the portrayed solutions that could be easily implemented in other platforms. As said before, while vendor-specific patterns provide detailed information on their composing elements, comprehensive of services, protocols, and parameters which should be used to implement them, agnostic patterns are much more general and should be applicable to different target platforms. However, they tend to provide too much vague information and are mostly limited to an architectural description of the portrayed solution. Similarities and differences exist between concepts defined in the different catalogs. For instance, comparing the definition of pattern composition expressed in the CloudPattern.org catalog and the description of cloud patterns offered by AWS, we can notice a strong similarity between the concept of mechanism from CloudPattern.org and the components used by Amazon patterns. Nevertheless, mechanisms are much more general and never give details about the actual implementation of the functionalities they offer, following the agnostic pattern trend. For example, the audit monitor mechanism specifies the necessity to instantiate a database to register accesses to the system (a log register), a login monitoring system, and an authentication service. All of these can be provided as smaller grain mechanisms or even patterns. Nothing is said about the character-istics of the database or of the access system. Despite their differences, among the presented catalogs we found patterns that aim at the same objectives offering similar solutions. Figure 2.11 presents some of the equivalences among these patterns and the catalogs to which they belong. The equivalences marked with an asterisk refer to the patterns that are not fully equivalent to each other but in which there are common components and aspects.

2.4.7 Semantic Cloud Patterns

While cloud patterns can be extremely useful to model cloud solutions and applica-tions and, therefore, can convey meaningful information to support software porting to the cloud and services' interoperation, they can be hampered by the lack of a shared machine-readable formalism for their representation. While for design patterns dif-ferent formalisms, based on the most disparate technologies, have been proposed and discussed [39], cloud patterns' formalisms are still missing. Works aimed at defining a semantic-based formalism for the accurate description of both static and behavioral aspects of cloud patterns can be found in [40, 41]. Here, cloud patterns' components are described using an OWL ontology, while the orchestration between such components is obtained through OWL-S.

2.4.8 Cloud Patterns: Engagement with Case Study and Positioning with Respect to Use Case Scenarios and Features

Using cloud patterns can ease the migration of an in-house complex application to the cloud and enhance interoperability between different platforms. For these reasons, the approach based on cloud patterns can be used in the use case scenarios *CSCC1* and *CSCC5*. Let us consider the example presented in Chap. 1: this represents a Business Intelligence application that can be roughly represented through the pipeline shown in Fig. 2.12. In particular, by comparing such pipeline with the provided example, it is possible to clearly map the ETL process that draws data from the CRM and ERP systems with the **data preprocessing** activities. Such activities are best represented through a **pipes and filters** pattern, presented and discussed in Sect. 2.4.3.

Each of the transformation activities can be executed through a cloud component owning computing capabilities, while communications can be dealt with through queue components. In Fig. 2.13, the correspondences existing between ETL steps (delimited by a blue square) and the filtering tasks defined by the pattern are high-lighted. At this point, it is clearly possible to instantiate a **filter** via a compute service offered either by Amazon through EC2 or Azure through a worker role (more on

cloudpatterns.org	cloud computing patterns. org	Cloud Pattern Amazon	Cloud Pattern Microsoft Azure
Dynamic Scalability / Elastic Resource Capacity	Elasticity Manager*	Scale-Up Pattern	Autoscaling Guidance + Throttling Pattern
Load Balanced Virtual Server Instances / Service Load Balancing	Elastic Load Balancer	Scale-Out Pattern	Autoscaling Guidance + Throttling Pattern
	Node based availability + Elastic Infrastructure	Multi Server Pattern	Compute partitioning Guidance*
Redundant Storage		Multi data-center Pattern	Multiple Datacenter Deployment Guidance
Persistent Virtual Network configuration		Floating IP Pattern	
RealTime Resource Availability		Deep Health Check Pattern	Health Endpoint Monitoring Pattern
	Managed Configuration	State Sharing Pattern	External Configuration Store Pattern
	Hybrid Processing	URL Rewriting Pattern	Static Content Hosting Pattern
Service State Management *		Cache Proxy Pattern	Cache-aside Pattern
	Hybrid Multimedia Web Application	Direct Hosting Pattern	Data Partitioning Guidance*
Resource Reservation *	Restricted Data Access Component	Private Distribution Pattern	Valet Key Pattern
Redundant Storage		DB Replication Pattern	Data Replication and Synchronization Guidance
	Strict Consistency	Read Replica Pattern	Command and Query Responsibility Segregation Pattern
Storage Workload Management		Sharding Write Pattern	Sharding Pattern
	Loose Coupling / Distributed Application	Queuing Chain Pattern	Pipes and Filters Pattern
Workload Distribution	Elastic Queue	Job Observer Pattern	Queue-based Load Leveling Pattern *
Broad Access	Three-Tier Cloud Application*	Multi Load Balancer Pattern	
Virtual Server Auto-Crash Recovery	Resiliency Management Process		Circuit Breaker Pattern

Fig. 2.11 Comparison between cloud patterns

Amazon and Azure in Sects. 4.1 and 4.5), while pipes can be instances of Azure Service Bus or AWS Simple Queue Service (SQS). Applying different patterns to each of the example, application's components can help developers to clearly understand their characteristics and thus correctly choose the cloud services needed to implement them, even on different platforms. In Fig. 2.13, two pipelines process data from different data sources, and services from independent providers, working at different service layers, are used.

Using semantically annotated cloud patterns, users can exploit the available knowledge base, containing meaningful relationships between patterns' participants and possible implementing cloud services, to automatically retrieve the cloud components needed to migrate their applications across cloud platforms or from on-premises infrastructure to new cloud environments. Also, the use of services offered by multiple providers would be automatically suggested and the information necessary for their interoperation (data and API formats, security protocols, and so on) would be provided accordingly. As an example, consider the SPARQL query shown in Listing 2.3: here a knowledge base, discussed in [40, 41] and containing information on cloud patterns and services, is used to retrieve the patterns' participants and all the services that can be used to implement them.

```
SELECT ?participant ?vendor ?component
WHERE {patternOntology:PipesAndFilters patternOntology
    :participant ? participant.
            ?participant patternOntology:equivalent+ ?
                component.
            ?component cloudServiceOntology:hasVendor ?
                vendor
    }
```

Listing 2.3 SPARQL query executed on OWL ontologies describing cloud patterns and services

Thanks to the "equivalent" property, defined in such ontology, it is possible to identify all the services which can be used to implement a specific pattern's participant, provided information on such equivalence has been previously included in the knowledge base.

The positioning of this approach with respect to our n-dimensional space is illustrated in Fig. 2.14.

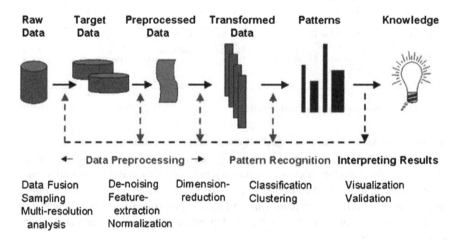

Fig. 2.12 Description of the data mining pipeline

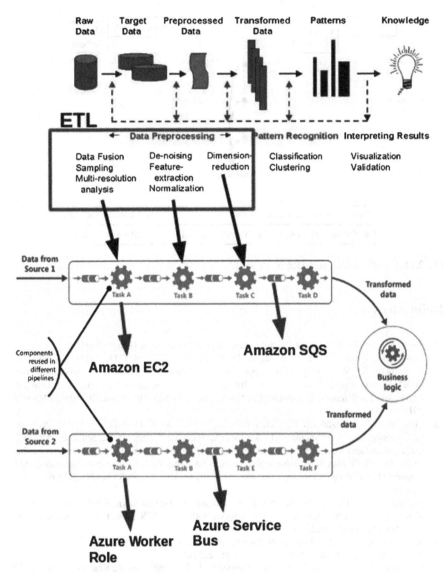

Fig. 2.13 Application design through cloud patterns

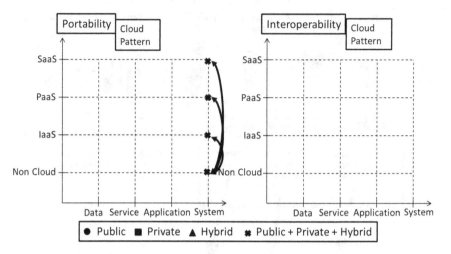

Fig. 2.14 Cloud patterns' approach positioning

References

1. Soley, R.: OMG: Model driven architecture. Object Management Group (2000)
2. Nitto, E.D., Silva, M.A.A.D., Ardagna, D., Casale, G., Craciun, C.D., Ferry, N., Muntes, V., Solberg, A.: Supporting the development and operation of multi-cloud applications: the MODAClouds approach. In: 2013 15th International Symposium on Symbolic and Numeric Algorithms for Scientific Computing (SYNASC), pp. 417–423 (2013). doi:10.1109/SYNASC.2013.61
3. Menychtas, A., Santzaridou, C., Kousiouris, G., Varvarigou, T., Orue-Echevarria, L., Alonso, J., Gorronogoitia, J., Bruneliere, H., Strauss, O., Senkova, T., Pellens, B., Stuer, P.: ARTIST methodology and framework: a novel approach for the migration of legacy software on the cloud. In: 2013 15th International Symposium on Symbolic and Numeric Algorithms for Scientific Computing (SYNASC), pp. 424–431 (2013). doi:10.1109/SYNASC.2013.62
4. http://www.paasage.eu
5. Mohagheghi, P., Berre, A.J., Sadovykh, A., Barbier, F., Benguria, G.: Reuse and migration of legacy systems to interoperable cloud services-the REMICS project. In: Proceedings of Mda4ServiceCloud '10 (2010)
6. MODAClouds project web-site. http://www.modaclouds.eu/
7. ARTIST project web-site. http://www.artist-project.eu/
8. Bergmayr, A., Bruneliere, H., Canovas Izquierdo, J., Gorronogoitia, J., Kousiouris, G., Kyriazis, D., Langer, P., Menychtas, A., Orue-Echevarria, L., Pezuela, C., Wimmer, M.: Migrating legacy software to the cloud with ARTIST. In: 2013 17th European Conference on Software Maintenance and Reengineering (CSMR), pp. 465–468 (2013). doi:10.1109/CSMR.2013.73
9. REMICS project web-site. http://www.remics.eu/
10. Sheth, A., Ranabahu, A.: Semantic modeling for cloud computing, part 2. IEEE Internet Comput. **14**(4), 81–84 (2010). doi:10.1109/MIC.2010.98
11. Aversa, R., Tasquier, L., Venticinque, S.: Management of cloud infrastructures through agents. In: 2012 Third International Conference on Emerging Intelligent Data and Web Technologies (EIDWT), pp. 46–53 IEEE (2012)
12. Mark, B., Jerry, H., Ora, L., Drew, M., Sheila, M., Srini, N., Massimo, P., Bijan, P., Terry, P., Evren, S., Naveen, S., Katia, S.: OWL-S: Semantic markup for web services. http://www.w3.org/Submission/2004/SUBM-OWL-S-20041122/

13. PrudHommeaux, E., Seaborne, A., et al.: SPARQL query language for RDF. W3C recommendation **15** (2008)
14. Horrocks, I., Patel-Schneider, P.F., Boley, H., Tabet, S., Grosof, B., Dean, M., et al.: SWRL: a semantic web rule language combining OWL and RuleML. W3C Member submission **21**, 79 (2004)
15. Singh, G., Kaur, N., Kaur, M.: Toward a unified ontology of cloud computing. Int. J. Comput. Technol. **3**(2) (2012)
16. Moscato, F., Aversa, R., Di Martino, B., Fortis, T., Munteanu, V.: An analysis of mOSAIC ontology for cloud resources annotation. In: 2011 Federated Conference on Computer Science and Information Systems (Fed-CSIS), pp. 973–980 IEEE (2011)
17. Di Martino, B., Cretella, G., Esposito, A.: Towards an unified OWL ontology of cloud vendors appliances and services at PaaS and SaaS level. In: Proceedings of the 8th International Conference on Computational Intelligence in Security for Information Systems (CISIS2014), pp. 570–575 (2014)
18. Di Martino, B., Cretella, G., Esposito, A., Carta, G.: Semantic representation of cloud services: a case study for openstack. In: Fortino, G., Di Fatta, G., Li, W.,Ochoa, S., Cuzzocrea, A., Pathan, M. (eds.) Internet and Distributed Computing Systems. Lecture Notes in Computer Science, vol. 8729, pp. 39–50. Springer International Publishing (2014)
19. Di Martino, B., Cretella, G., Esposito, A., Sperandeo, R.G.: Semantic representation of cloud services: a case study for Microsoft Windows Azure. In: Proceedings of the 6th International Conference on Intelligent Networking and Collaborative Systems (INCoS-2014), pp. 647–652 (2014)
20. mOSAIC project web-site. http://www.mosaic-cloud.eu/
21. Cretella, G., Di Martino, B.: Towards a semantic engine for cloud applications development. In: Proceedings—2012 6th International Conference on Complex, Intelligent, and Software Intensive Systems, CISIS 2012, pp. 198–203 (2012). doi:10.1109/CISIS.2012.159
22. Cretella, G., Di Martino, B.: Semantic and matchmaking technologies for discovering, mapping and aligning cloud providers's services. In: Proceedings of the 15th International Conference on Information Integration and Web-based Applications and Services (iiWAS2013), pp. 380–384 (2013)
23. Cloud4SOA project web-site. http://www.cloud4soa.eu/
24. Trajanov, D., Stojanov, R., Jovanovik, M., Zdraveski, V., Ristoski, P., Georgiev, M., Filiposka, S.: Semantic sky: a platform for cloud service integration based on semantic web technologies. In: Proceedings of the 8th International Conference on Semantic Systems, pp. 109–116. ACM (2012)
25. Talia, D.: Cloud computing and software agents: towards cloud intelligent services. In: WOA, vol. 11, pp. 2–6. Citeseer (2011)
26. Venticinque, S., Tasquier, L., Di Martino, B.: Agents based cloud computing interface for resource provisioning and management. In: 2012 Sixth International Conference on Complex, Intelligent and Software Intensive Systems (CISIS), pp. 249–256 (2012). doi:10.1109/CISIS. 2012.139
27. Amato, A., Tasquier, L., Copie, A.: Vendor agents for IAAS cloud interoperability. In: Intelligent Distributed Computing VI, pp. 271–280. Springer (2013)
28. Metsch, T., Edmonds, A., et al.: Open cloud computing interface-infrastructure. In: Standards Track, no. GFD-R in The Open Grid Forum Document Series, Open Cloud Computing Interface (OCCI) Working Group, Muncie (IN) (2010)
29. Petcu, D., Martino, B., Venticinque, S., Rak, M., Mhr, T., Lopez, G., Brito, F., Cossu, R., Stopar, M., Perka, S., Stankovski, V.: Experiences in building a mOSAIC of clouds. J. Cloud Comput. **2**(1), 12 (2013). doi:10.1186/2192-113X-2-12. http://dx.doi.org/10.1186/2192-113X-2-12
30. Sim, K.M.: Agent-based cloud commerce. In: 2009 IEEE International Conference on Industrial Engineering and Engineering Management, IEEM, pp. 717–721. IEEE (2009)
31. Iannucci, P., Gupta, M., et al.: IBM SmartCloud: Building a Cloud Enabled Data Center. IBM Redbooks, New York (2013)
32. AWS cloud design patterns. http://en.clouddesignpattern.org

33. Windows Azure cloud design patterns. http://msdn.microsoft.com/en-us/library/dn568099. aspx
34. Cloud patterns. http://cloudpatterns.org
35. Fehling, C., Retter, R.: Cloud Computing Patterns. University of Stuttgart, Stuttgart (2011)
36. IBM: A design pattern definition. http://expertintegratedsystemsblog.com/index.php/2012/07/getting-back-to-the-basics-what-is-a-pattern/
37. Brandle, C., Grose, V., Hong, M.Y., Imholz, J., Kaggali, P., Mantegazza, M., et al.: Cloud Computing Patterns of Expertise. IBM Redbooks, New York (2014)
38. Mell, P., Grance, T.: The NIST Definition of Cloud Computing. Computer Security Division, Information Technology Laboratory, National Institute of Standards and Technology, Gaithersburg (2011)
39. Taibi, T.: Design Pattern Formalization Techniques. IGI Publishing, Hershey (2007)
40. Di Martino, B., Cretella, G., Esposito, A.: Semantic and agnostic representation of cloud patterns for cloud interoperability and portability. In: Proceedings of the IEEE Fifth International Conference on Cloud Computing Technology and Science (CloudCom2013) (2013)
41. Di Martino, B., Esposito, A.: Towards a common semantic representation of design and cloud patterns. In: Proceedings of International Conference on Information Integration and Web-Based Applications and Services, p. 385. ACM (2013)

Chapter 3
Cross-Platform Cloud APIs

3.1 Introduction to Cross-Platform Cloud APIs

Cross-platform cloud APIs are high-level APIs that provide access to multiple cloud platforms through a common, shared interface. They enable an abstract description of the services and functionalities exposed by a set of target platforms, providing a uniform vision of their offers. Using a cross-platform, API can bring benefits for both portability and interoperability:

- Applications using cross-platform APIs can be easily ported across multiple platforms, as long as such platforms are supported by the exploited APIs. In general, a developer would have only to update login information in order to let her application access services on a new platform but, other than that, no other changes would be necessary.
- Cloud platforms supported by a cross-platform API could communicate and exchange information using such a shared interface, thus enhancing their interoperability.

When it comes to working principles, cross-platform APIs are generally implemented either as wrappers or as adapters.

- **Wrappers** completely hide the actual API calls to a target platform: consumers see the cross-platform APIs only, and they can use them independently from the actual vendor involved since the wrapper automatically translates the issued calls. In many cases, the term **Driver** can be used interchangeably.
- **Adapters** act as translators between API calls without imposing a particular interface to a client: a user issues an API call using an interface she knows well and the adapter translates it to the correct API call for the target platform. Obviously, both the interface used by the client and the one exposed by the target platform must be supported by the cross-platform API. **Connector** is another term describing the same approach.

© The Author(s) 2015
B. Di Martino et al., *Cloud Portability and Interoperability*,
SpringerBriefs in Computer Science, DOI 10.1007/978-3-319-13701-8_3

As regards the technologies and programming languages used to expose cross-platform APIs, we find the same ones used for single platform APIs. Generally, libraries written in Java, Python, Ruby, and other languages are available to consumers who can include them in their applications to use the shared API. Also, such APIs can be exposed through a REST interface or described in WSDL.

3.2 DeltaCloud

DeltaCloud [1] is an open-source Apache project which aims at defining a REST-based API to access any kind of cloud platform exposing its services at the IaaS level. Written entirely in Ruby (but with the possibility to exploit client libraries to use different languages), it offers the user the opportunity to leverage *classic* Delta-Cloud, DMTF, CIMI (see Sect. 5.3), and EC2 APIs (see Sect. 4.1) representing the DeltaCloud abstraction API, to manage different cloud platforms. This API works as a wrapper around these platforms, abstracting and hiding their differences. For every cloud provider there is a driver that is able to interpret the cloud provider's native API, so that the user does not need to deal with the specific API directly. This also means that, if a particular API published by a provider was updated and changed, the user program would not be affected because DeltaCloud specifications are guaranteed to remain stable. The user could also develop an application having a specific platform in mind and then decide to migrate to another, Delta-Cloud compatible, one. A list of the supported providers and of the relative APIs is provided online and it includes many cloud platforms such as Amazon EC2, IBM SmartCloud, GoGrid, and OpenStack (see Sect. 4.2). Being an open-source project, contributors can freely add their custom drivers to the list and modify the existing ones.

3.2.1 How to Use DeltaCloud

DeltaCloud developers have provided users with a very simple means to leverage their abstract API or the supported CIMI and EC2 standards. In particular, invocation of such APIs can be done through an HTTP client, which can be represented by a Web browser, the provided DeltaCloud **Ruby** client, or even a custom client. The RESTful nature of the APIs makes them extremely flexible and easy to invoke. Figure 3.1 shows how a user can connect to a cloud provider using either DeltaCloud or CIMI APIs through an HTTP client. The client connects to a DeltaCloud instance, which can be a public one or simply running on a local machine or a private server. The user just needs the credentials to access such an instance (location is not important). The DeltaCloud instance selects the correct driver to use, according to the information received from the client on the target cloud platform, which wraps the API invocation in order to communicate with the selected cloud provider.

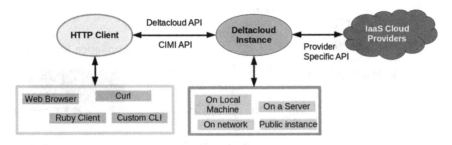

Fig. 3.1 DeltaCloud usage

3.3 OpenNebula

OpenNebula [2, 3] is a cloud computing toolkit that enables consumers to manage heterogeneous and distributed data-center infrastructures, specifically addressing IaaS platforms and focusing on virtualized infrastructures in data centers or clusters (private cloud). Support to hybrid solutions, which combine local and public cloud infrastructures, is also provided together with interfaces to expose public cloud's functionalities for virtual machine, storage, and network management. Announced collaborations with Windows Azure and SoftLayer will improve the support to hybrid cloud solutions. Apart from a set of native APIs, offered via XML-RPC and Java or Ruby bindings, OpenNebula also implements Amazon EC2, OGF OCCI, and vCloud APIs. Interoperability is supported by leveraging and implementing existing standards, which leads to the development of adapters and transformers for APIs provided by different cloud providers. Remarkable is the possibility to exploit adapters for DeltaCloud (Sect. 3.2) and Libcloud (Sect. 3.6), a standard client library for many popular cloud providers, written in Python.

3.3.1 Different Users' Perspectives

One of the strongest points of OpenNebula is the extreme flexibility of their exposed interface and functionalities, which can be adapted to meet the needs of the different API users. There are four different perspectives to interact with OpenNebula. The **cloud consumers'** perspective provides interfaces such as OCCI, EC2 Query, and EBS, together with a cloud user view of the **Sunstone** dashboard. Functionalities offered to consumers include: image catalogs, network catalogs, VM template catalog, virtual resource control and monitoring features, and multitier cloud application control and monitoring capabilities.

Cloud advanced users and operators can leverage administration-oriented interfaces such as a Unix-like command line interface (CLI) and the powerful Sunstone GUI (as already stated, different views are provided to consumers and operators). The pool of functionalities and management components that the administrators can

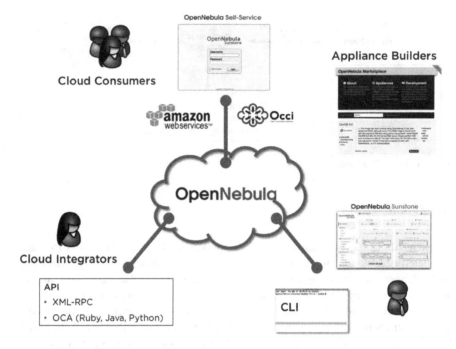

Fig. 3.2 Overview of OpenNebula interfaces

choose from is quite rich and includes management of users and groups, virtualization support, host management capabilities, monitoring and networking functionalities, storage capabilities, security mechanisms, high performance support, multiple zones management, support to cloud bursting, and App Market.

Cloud integrators can leverage extensible low-level APIs for Ruby, Java, and XML-RPC. **Appliance builders** have full access to the App Marketplace, containing a catalog of virtual appliances ready to run in OpenNebula environments. An overview of the different available interfaces is provided in Fig. 3.2, extracted from [2].

3.3.2 OpenNebula Architecture

OpenNebula architecture, visually described in Fig. 3.3, is organized in layers. The upper layer contains tools used to interact with an OpenNebula instance and manage virtual machines. The **scheduler** tool is a component that automatically searches for physical hosts to deploy newly defined VMs. The **command line interface** is used to issue commands to manage the entire life cycle of instantiated VMs. The middle layer contains the core components of the OpenNebula architecture.

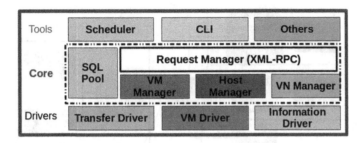

Fig. 3.3 Overview of OpenNebula architecture

- The **Request Manager** provides an XML-RPC interface to manage and gets information about "ONE" entities.
- The **SQL Pool** is the database that holds the state of "ONE" entities.
- The **Virtual Machine Manager** takes care of the entire VM lifecycle.
- The **Host Manager** holds the information about hosts and knows how to interact with them.
- The **Virtual Network Manager** component is in charge of generating MAC and IP addresses.

The bottom layer of the architecture contains the drivers used to communicate with cloud platforms at different levels.

- **Transfer Drivers** take care of image processing such as cloning, deleting, and creating swap locations.
- **Virtual Machine Drivers** concretely manage the life cycle of a virtual machine by translating commands issued through the CLI in calls understandable from the VM. Responsibilities of such drivers comprehend deploying, shutting down, polling, and migration of VMs.
- **Information Drivers** execute scripts in physical hosts to gather information about them such as total memory, free memory, total CPUs, CPU consumed, and so on.

3.4 DeltaCloud and OpenNebula: Engagement with Case Study and Positioning with Respect to Use Case Scenarios and Features

As further stated in Sect. 3.8, most of the existing cross-platform APIs support IaaS functionalities only. So, an implementation of the example Business Intelligence application provided in Sect. 1.2.2 could be possible only if IaaS resources from supported providers were used. Figure 3.4 shows how it could be possible to map the different components of our example application, which have been delimited

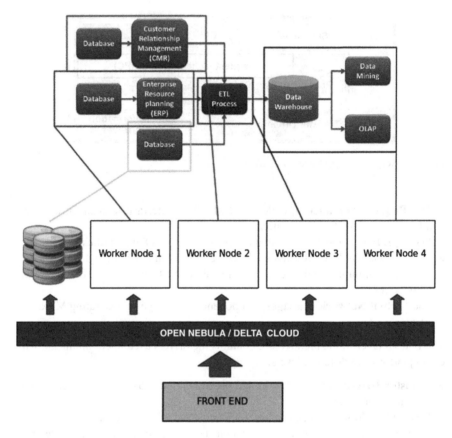

Fig. 3.4 Possible implementation of the Business Intelligence application through cross-platform APIs

with frames, to IaaS services offering computing and storage capabilities, here simply denoted as **Worker Nodes**. In our example, we suppose that the single application components are independent applications themselves, running either on different or on the same software systems within an on-premise environment, thus sharing resources and communicating with each other. In this case, the system(s) they run on could be encased in a VM and then be deployed elsewhere: resources necessary for their execution would be provided by the target cloud vendor(s), while communication could be taken care of by the cross-platform APIs. In the end, we could find one of the three different situations:

• Only some of the application components, such as the CRM software and its related storage, are actually migrated to the cloud, thus realizing a hybrid interoperability scenario.

- After the migration, some of the application components could be running within VMs hosted on different cloud platforms, interoperating at the service level through the shared APIs.
- At a later time, the owner of the Business Application could decide to move some of its components (encased in one or more VMs) to another cloud platform: thanks to the shared APIs, this would be possible without modifying the application's code (of course, some configuration changes would be needed). This would enable application portability.

Because of the lack of PaaS API support, the single application components cannot be entirely reimplemented for the target platform without the risk of its "lock-in", making it difficult to remigrate them to another platform at a later time.

These two solutions address the use case scenario CSCC S1 since they enable portability among the supported infrastructures. The key aspect of these solutions is of course the interoperability at the IaaS level. It is possible to use a single API to access compute resources and storages hosted by different cloud providers and for this reason these solutions address the use case scenario SCSS S2 only at the IaaS level. In the context of the use case scenario CSCC S4, OpenNebula would enable the communication between the private subsystem and the cloudified part of the application. In the context of the use case scenario CSCC S5, the use of OpenNebula or DeltaCloud will enable the migration of non-cloud applications to the IaaS solution. For what concerns application portability, the use of OpenNebula or DeltaCloud will enable an automatic portability process among the supported IaaS platforms. The interoperability features offered by these solutions consist in the transparent usage of resources offered by different cloud providers and accessible through a common API. The positioning of these solutions according to our n-dimensional space is illustrated in Fig. 3.5.

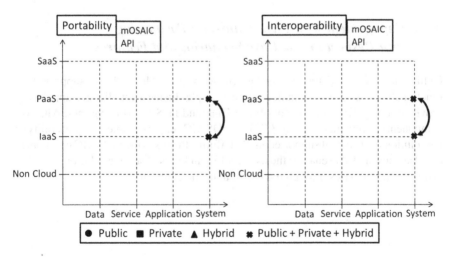

Fig. 3.5 Cross-platform API solution positioning

3.5 mOSAIC API

The main aim of the mOSAIC project is to offer a solution for application portability and interoperability across multiple clouds [4]. To achieve this objective, the mOSAIC platform offers first of all an API, implemented in Python, Java, and Erlang, to develop components that run on top of its platform. mOSAIC's API is designed to be event-driven, and the communication among mOSAIC components takes place through message queues. The mOSAIC's basic component is the **cloudlet**, an event-driven and stateless component whose functionalities do not depend on the number of its instances at runtime (has a degree of autonomy). The cloudlets are able to access cloud services through **connectors**. The concept of a connector is introduced to ensure the independence from the cloud service interfaces. A connector is a concrete class that abstracts the access to cloud resources and defines the set of events to which the cloudlet should react. The connectors access cloud services using **drivers**, which actually implement the cloud service interfaces. They can be interpreted as wrappers of native resource APIs or uniform APIs, like OpenStack. These wrappers are able to send and receive messages from the mOSAIC's message queues. Until now, the mOSAIC software repository includes modules for more than ten public clouds. Among these, mOSAIC supports well-known providers like Amazon, Rackspace, and GoGrid, as well as European cloud providers including Flexiant (UK), CloudSigma (Switzerland), NIIFI (Hungary), Arctur and Hostko (Slovenia), the last two using VMware vCloud. Moreover, private clouds are built using open-source technologies like Eucalyptus, OpenNebula, CloudStack, and the already cited OpenStack.

Figure 3.6 reports mOSAIC's stack showing the components of the integration platform: the already cited APIs belong to the **application support** level, on which mOSAIC's IaaS and PaaS functionalities are based.

3.5.1 Engagement with Case Study and Positioning with Respect to Use Case Scenarios and Features

Unlike most of the existing cross-platform APIs, mOSAIC API also supports IaaS functionalities. An implementation of the example Business Intelligence application provided in Fig. 1.6 could be possible at PaaS and IaaS levels using mOSAIC API components as illustrated in Fig. 3.7. The mOSAIC support enables the portability of the solution over several supported IaaS platforms thanks to the mOSAIC component cloud agency, and also enables the use of different kinds of storage solutions for SQL and NoSQL databases through the use of agnostic connectors.

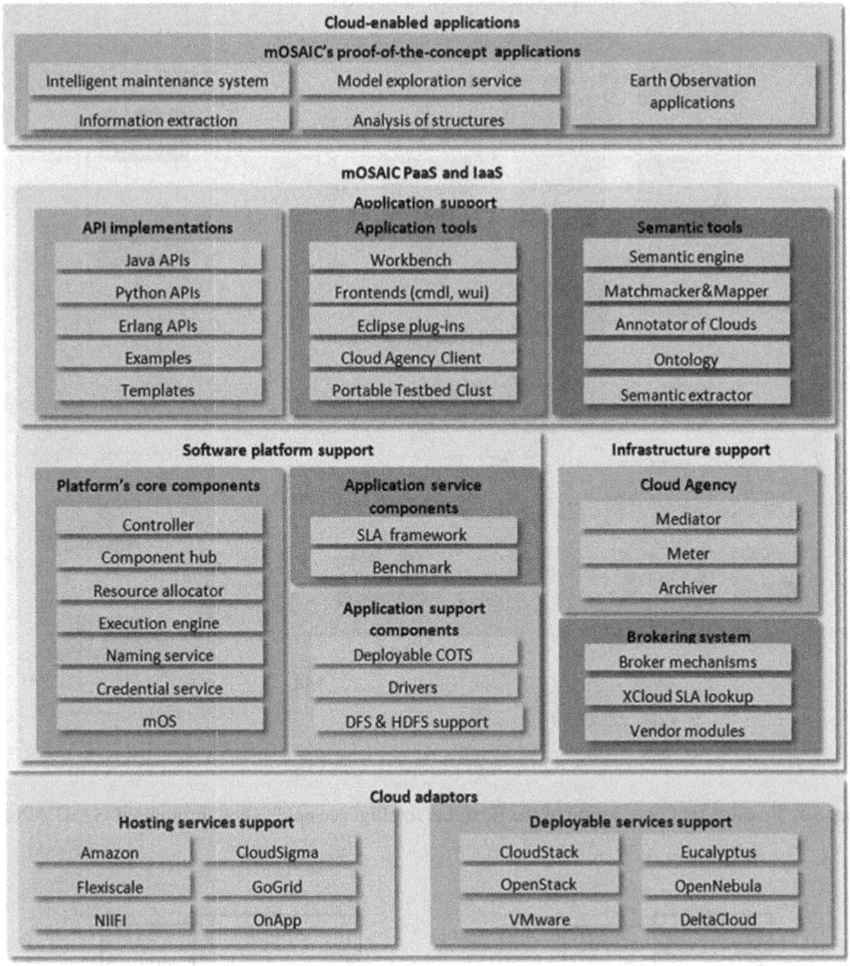

Cloud-enabled applications		
mOSAIC's proof-of-the-concept applications		
Intelligent maintenance system	Model exploration service	Earth Observation applications
Information extraction	Analysis of structures	

mOSAIC PaaS and IaaS

Application support

API implementations	Application tools	Semantic tools
Java APIs	Workbench	Semantic engine
Python APIs	Frontends (cmdl, wui)	Matchmacker&Mapper
Erlang APIs	Eclipse plug-ins	Annotator of Clouds
Examples	Cloud Agency Client	Ontology
Templates	Portable Testbed Clust	Semantic extractor

Software platform support — **Infrastructure support**

Platform's core components	Application service components	Cloud Agency
Controller	SLA framework	Mediator
Component hub	Benchmark	Meter
Resource allocator		Archiver
Execution engine	Application support components	Brokering system
Naming service	Deployable COTS	Broker mechanisms
Credential service	Drivers	XCloud SLA lookup
mOS	DFS & HDFS support	Vendor modules

Cloud adaptors

Hosting services support		Deployable services support	
Amazon	CloudSigma	CloudStack	Eucalyptus
Flexiscale	GoGrid	OpenStack	OpenNebula
NIIFI	OnApp	VMware	DeltaCloud

Fig. 3.6 mOSAIC stack

mOSAIC addresses the use case scenario CSCC S1 since it enables portability among the supported infrastructures at the IaaS level and between the supported services at the PaaS level. The interoperability is possible thanks to the single API to access compute resources and storages hosted by different cloud providers and for this reason this solution addresses the use case scenario SCSS S2 at the IaaS and PaaS levels. The positioning of these solutions according to our n-dimensional space is illustrated in Fig. 3.8.

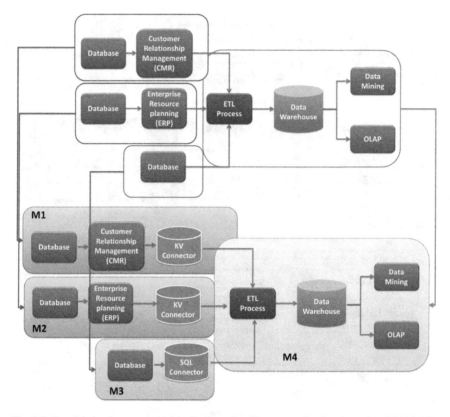

Fig. 3.7 Possible implementation of the Business Intelligence application through mOSAIC APIs

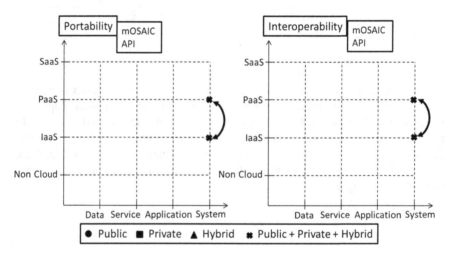

Fig. 3.8 mOSAIC API solution positioning

3.6 Apache Libcloud

Libcloud [5] is a library, entirely written in Python, allowing interactions among several popular cloud service providers. In particular, Libcloud consists of a unified API that hides differences between different providers thanks to a high-level abstraction of the services and APIs exposed by the different providers. Its main objective is to simplify developers' efforts at building products that are able to work with services coming from different cloud vendors, thus easing interoperability issues. The number of supported providers and functionalities is quite interesting. A complete list of such providers can be found at [6]. Libcloud API allows the management of resources belonging to four main categories:

- **Cloud Servers and Block Storage** such as Amazon EC2 and Rackspace cloud-servers;
- **Cloud Object Storage and CDN**, among which are Amazon S3 and Rackspace CloudFiles;
- **Load Balancers as a Service**, including Amazon Elastic Load Balancer and GoGrid LoadBalancers;
- **DNS as a Service** such as Amazon Route 53 and Zerigo;

For all cited Amazon services, see Sect. 4.1. Each of the categories listed above is managed through a specific Libcloud component.

3.7 Apache JClouds

The JClouds' [7] API shares many features and concepts with those of the Libcloud API, both having the same objective to define a high-level API to access and manage services provided by different cloud vendors in a uniform manner. In particular, JClouds offers an open-source library, operable through Java or Clojure [8], enabling the creation of applications that are portable across different cloud platforms, giving the user the possibility to still use cloud-specific features. The list of supported providers includes important players such as Amazon, Azure, and OpenStack. Using Java and Clojure libraries, users can interact with services offered by cloud providers without having to deal directly with REST APIs and, at the same time, using an abstract representation of the target resources. In particular, the library offers to manage resources included in three different categories:

- **ComputeService** The portable compute interface allows users to provision their infrastructure in any supported cloud provider. The entire life cycle of a compute resource, from deployment configuration, provisioning, and bootstrapping, is manageable through JClouds. The API offers the possibility to start multiple machines at once, thanks to the available configuration templates that specify configuration properties like: the *Image* used to create the computing node instance; the *Hardware* on which the instance will run, comprehensive of CPU speed, available RAM, and disk space; the *Location* where the machine will run (in Amazon, the

Region and Availability Zone, see Sect. 4.1). Installing software and run deployment scripts at instantiation time is also possible.

- **BlobStore** simplifies the management of key-value storages offered by different providers. It offers both asynchronous and synchronous APIs, as well as Map-based access to data.
- **LoadBalancer** provides a common interface to configure load balancers in any cloud platform that supports them. In order to operate with a load balancer, a user just has to create an instance of it on the target platform and then to attach computing nodes to it. The LoadBalancer functionalities are still in beta release at the time of writing.

Table 3.1 Summary of the presented cross-platform APIs' capabilities

API	Accessible through	Supported languages	Supported platform/standard	Type
DeltaCloud	REST interface, Ruby Client (Curl), C/C++ Library	Ruby, C/C++	Amazon EC2, Eucalyptus, Fujitsu FGCP, IBM SmartCloud, GoGrid, OpenNebula, Rackspace, RHEV-M, RimuHosting, Terremark, vSphere, DigitalOcean, ArubaCloud, OpenStack, ProfitBricks	Wrapper
OpenNebula	XML-RPC	Java, Ruby, Python	Amazon EC2, OCCI	Adapter
LibCloud	Python Library	Python	Abiquo, Bluebox Blocks, Brightbox, CloudFrames, CloudSigma (API v2.0), CloudStack, DigitalOcean, DreamHost, Amazon EC2, Enomaly Elastic Computing Platform, ElasticHosts, Eucalyptus, Exoscale, Gandi, Google Compute Engine, GoGrid, HostVirtual, HP Helion Public Cloud, IBM SmartCloud Enterprise, Ikoula, Joyent, Kili Public Cloud, KTUCloud, Libvirt, Linode, NephoScale, Nimbus, Ninefold, OpenNebula (v3.8), OpenStack, OpSource, Outscale INC, Outscale SAS, Rackspace Cloud, RimuHosting, ServerLove, SkaliCloud, SoftLayer, vCloud, VCL, vCloud, Voxel, VoxCLOUD, VPS.net, VMware vSphere	Wrapper
JClouds	Java Library	Java	AWS, Bluelock, CloudSigma, DigitalOcean, Docker, ElasticHosts, Go2Cloud, GoGrid, Google Compute Engine, Green House Data, vCloud, HP, Ninefold, Open Hosting, Rackspace, ServerLove, SkaliCloud, SoftLayer	Wrapper
mOSAIC	Java Library	Java, Erlang, Python	Amazon, GoGrid, Rackspace, Flexiant, CloudSigma, NIIFI, Arctur, Hostko, Eucalyptus, OpenNebula, CloudStack, OpenStack	Adapter

3.8 Comparative Analysis

Table 3.1 reported in this section summarizes the different characteristics of various APIs introduced in this chapter. Some of such APIs, like LibCloud and JCloud, support a wide range of platforms, even if they are limited to specific services. In general, wrapper APIs accessible through libraries which can be directly imported by developers in their projects offer a wider support in terms of manageable services and platforms. The possibility to extend the offered APIs in order to include new platforms and services is a common characteristic, and developers are encouraged to contribute. This is also endorsed by the open-source license that characterizes these APIs. Despite the good number of platforms and services managed by the different cross-platform APIs, the general tendency is to support the IaaS only. The PaaS and SaaS offers are not contemplated yet by these APIs, even though a series of projects have been trying to extend the existing libraries. This is probably due to the lack of a mature standard for the PaaS interfaces, to which such cross-platform APIs could adhere, while a plethora of solutions exist for IaaS (see Chap. 5 for a list of mature and emerging standards) even if they are not always adopted by vendors.

References

1. Deltacloud. https://deltacloud.apache.org/
2. Opennebula web-site. https://opennebula.org/
3. Moreno-Vozmediano, R., Montero, R.S., Llorente, I.M.: IaaS cloud architecture: from virtualized datacenters to federated cloud infrastructures. Computer **45**(12), 65–72 (2012)
4. Petcu, D., Di Martino, B., Venticinque, S., Rak, M., Máhr, T., Lopez, G.E., Brito, F., Cossu, R., Stopar, M., Šperka, S., et al. Experiences in building a mOSAIC of clouds. J. Cloud Comput. **2**(1), 1–22 (2013)
5. Libcloud reference web-site. https://libcloud.apache.org/index.html
6. Libcloud documentation. https://libcloud.readthedocs.org/en/latest/
7. The Java multi-cloud toolkit. https://jclouds.apache.org/
8. Clojure. http://clojure.org/

Chapter 4
Ready-to-Go Solutions

4.1 Amazon Web Services (AWS)

Amazon Web Services (AWS) [1] is a comprehensive cloud services platform that offers compute power, storage, network, content delivery, and other cloud-based functionalities. The number of services exposed by AWS is impressive and describing each of them here would be impossible, so we focus on the most popular ones. Figure 4.1 lists AWS by grouping them into different categories.

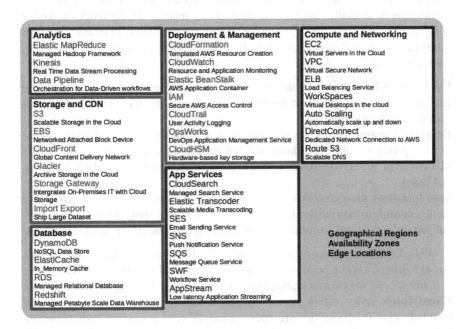

Fig. 4.1 List of cloud services provided by AWS

© The Author(s) 2015
B. Di Martino et al., *Cloud Portability and Interoperability*,
SpringerBriefs in Computer Science, DOI 10.1007/978-3-319-13701-8_4

Access to AWS is granted through different means. The **AWS Management Console** is a Web-based GUI enabling management of all AWS resources, from EC2 instances to DynamoDB tables. Through this interface, consumers can schedule and execute different tasks comprehending instances' management and applications' deployment.

Besides the web graphical interface, users can also utilize the **command line interface** (CLI) provided by Amazon. Based on Python, such an interface enables complete control, management, and configuration capabilities over AWS instances. The main benefit from using the CLI comes from the possibility to use custom scripts to control AWS instances.

AWS APIs are available to control instances either through HTTP messages (using a RESTful style) or by means of SDKs, enabling the use of a tailored API for a chosen programming language or platform.

4.1.1 Compute Services

The most popular services delivered by AWS are surely represented by **Amazon Elastic Compute Cloud** (EC2), **Auto-Scaling**, and **Elastic Load Balancing**. These services, often included in developers' solutions and suggested in many of the Amazon design patterns (see Sect. 2.4.4), are generally used together to deliver compute capabilities in an easily scalable and manageable manner. EC2 provides resizable compute capacity in the cloud, giving consumers the opportunity to leverage the IaaS infrastructure offered by Amazon at its full. Automatically scalable, EC2 resources are offered in different flavors:

On-Demand Instances incarnate the cloud philosophy of "You pay for what You get", since instances automatically scale with requests and payments are based on actual consumption only.

Reserved Instances let users reserve a certain amount of resources for a fixed period of time, giving also the opportunity to resell unused instances (through the Amazon Marketplace). Reserved instances come in three different dimensions: light, medium, and heavy.

Spot Instances allow customers to bid on unused Amazon EC2 capacity, paying for them according to their current price. They are useful when applications can run without particular time and performance constraints, but at generally much lower prices.

While EC2 instances are natively capable to scale according to the actual compute power requests, it is also possible to associate the auto-scaling service to them in order to better specify the custom scaling criteria and policies. Auto-scaling is part of the **CloudWatch** service, offering monitoring capabilities on EC2 instances. The elastic load balancing (ELB) service provides the capability to homogeneously distribute traffic among EC2 instances, in order not to overwhelm a single compute node with requests and avoid bottlenecks. Also, it detects faulty instances and redirects traffic through healthy ones, when available. ELB can distribute workloads within the same **Availability Zone** (see [2] for more details) or across different ones.

4.1.2 Storage and Database Services

AWS offers several storage services, with peculiar capabilities and objectives. **Simple storage service** (S3) is probably one of the most known and supported storage offers in the whole cloud environment. The service provides storage space for data of any kind and any dimension, organized in practical structures known as "Buckets". High data availability and reliability have made this service extremely popular among developers. **Glacier** is another storage offer from AWS, which is focused on archiving and backing-up infrequently used data at very low prices. The counterpart is represented by the slow access and retrieve time this kind of storage offers, making it unsuitable for applications consuming and/or producing data at a fast pace. **Elastic block storage** (EBS) provides persistent block level storage volumes for use with Amazon EC2 instances. EBS instances can be either exposed as devices within an EC2 instance, or they can be attached to and accessed through a network.

Apart from raw storage capabilities, AWS also offers different databases as service functionalities, mainly represented by:

- **Relational Database Service** (RDS), providing access and management functionalities for a relational database completely delivered in the cloud. RDS is compliant with popular relational database solutions, including MySQL, Oracle, and PostgreSQL.
- **DynamoDB** is a fast NoSQL database service, organized in easily accessible and retrievable tables. In order to ensure availability and durability, data are stored on Solid State Drives (SSDs) replicated across three availability zones.
- **ElastiCache** is a Web service offering in-memory cache in the cloud. The service is meant to improve the performance of Web applications by allowing you to retrieve information from fast, managed, in-memory caches, instead of relying entirely on slower disk-based databases. The service currently supports two open-source in-cache solutions represented by: **Memcached**, a memory object caching system; **Redis**, an in-memory key-value store supporting data structures such as sorted sets and lists.

4.1.3 Networking Services

AWS offers network-related services to easily connect and enable communications among Amazon compute instances and storages. In particular, the **Route 53** service provides Domain Name System functionalities, allowing connections between services running in AWS, such as an EC2 instance, an elastic load balancer, or an Amazon S3 bucket. Route 53 can also be used to route users to infrastructures outside of AWS. Using Amazon **identity and access management** (IAM), it is possible to control who is able to modify routing tables in AWS. **Direct connect** makes it possible to establish a dedicated network connection from on-premise infrastructures to AWS.

4.1.4 Deployment and Management

Amazon **CloudWatch** is a monitoring service enabling creation of metrics, policies, log files, and alarms to have complete control of AWS resources or of applications running on them. CloudWatch can be used together with other services, such as auto-scaling, to provide workload balancing functionalities and respond to fluctuating compute power requests. AWS **Elastic Beanstalk** allows users to easily deploy and scale Web applications and services developed with several programming languages and frameworks such as Java, .NET, PHP, Node.js, Python, Ruby, and Docker. Target servers include such popular products as Apache HTTP Server, Apache Tomcat, Nginx, Passenger, and IIS 7.5/8. Users just have to upload their code, and Elastic Beanstalk automatically executes all the operations needed to correctly launch the applications, from resource provisioning to health monitoring. **CloudFormation** offers developers and system administrators a means to completely control and manage the workflow of a collection of AWS resources, including their provisioning, running, and updating. CloudFormation offers a set of predefined templates describing relationships existing between resources and their workflow, but users can create custom ones.

A list of available templates is available (at the time of writing) at [3].

4.1.5 Engagement with Case Study and Positioning with Respect to Use Case Scenarios and Features

Amazon offers a huge variety of cloud services that allows to deploy the entire business application on their cloud. A case of how our example application can be implemented using Amazon cloud services is illustrated in Fig. 4.2. Some components such as the CRM and ERP are provided at the SaaS level, but can also be deployed through VMs running on Amazon EC2 instances. Amazon Redshift, specifically designed for OLAP, would be used as the data warehouse, while Amazon RDS or DynamoDB offers the database component; the CRM and ERP components are not directly offered by Amazon, but several SaaS solutions can be automatically deployed in the Amazon IaaS via the AWS marketplace. Useful for the ETL process is the Amazon EMR (Elastic MapReduce) service alongside third-party tools for data analysis and extraction. Since all components of the application can be deployed in AWS, or they can be substituted by third-party SaaS services able to interoperate, use cases CSCC5 and CCUC1 are supported. Amazon offers the support for interoperability not only with AWS products. For instance, Amazon RDS allows to deploy multiple editions of the Oracle Database and it is also possible to build an OpenShift platform on Amazon EC2: CSCC2 use case is thus supported at both the IaaS and PaaS levels. Also, since VMs can be transferred to and from AWS using the available APIs, as long as the right format is used (sometimes a conversion could be necessary), use case CCUC3 is partially supported. The positioning of the Amazon solution over our n-dimensional space is illustrated in Fig. 4.3.

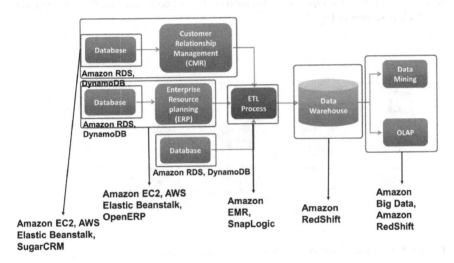

Fig. 4.2 Application design through Amazon AWS

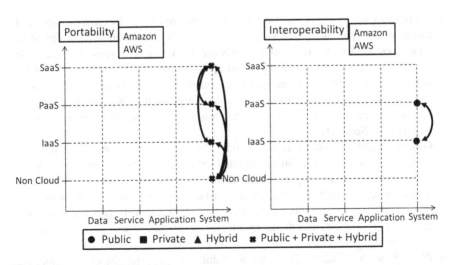

Fig. 4.3 Amazon solution positioning

4.2 OpenStack

OpenStack [4] is an open-source cloud operating system that provides compute, storage, and networking resources together with virtualization technologies fully accessible and manageable through a set of powerful APIs, command line interface (CLI) tools, or software development kits (SDKs) provided for different programming languages. It is also possible to leverage a graphical dashboard that allows users to manage and monitor resources provided by the platform. Services are offered at an IaaS level. The current stable OpenStack release is named IceHouse, which was released in April 2014. OpenStack is based on a set of core services, the organization of which is shown in Fig. 4.4:

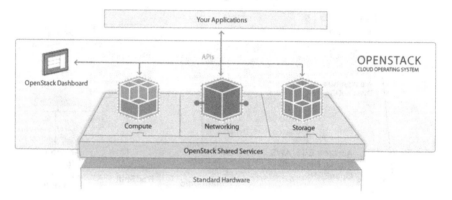

Fig. 4.4 OpenStack services organization. Image taken from [4]

- **Compute** (Nova) offers on-demand computing resources by provisioning and managing large networks of virtual machines. The architecture is designed to scale horizontally on the standard hardware. Compute resources are accessible via APIs for developers building cloud applications and via Web interfaces for administrators and users. When launching computing instance on OpenStack, users can exploit default configurations to reserve a predefined amount of resources for their virtual machines to run. Such configurations resemble Amazon flavors described in Sect. 4.1.1.
- **Object Storage** (Swift) provides a fully distributed storage platform for data backup and archiving; it is fully integrated with applications running in the Open-Stack environment and expands external storage of compute instances. All objects stored are replicated thrice in as-unique-as-possible zones, which can be defined as a group of drives, a node, a rack, etc. The service supports dynamical creation and deletion of nodes and disks, which can be substituted and swapped with no downtime.
- **Block Storage** (Cinder) allows the availability of persistent block level storage devices, directly connected with OpenStack compute instances. Block storage

allows block devices to be exposed and connected to compute instances for expanded storage, better performance, and integration with enterprise storage platforms.

- **Networking** (Neutron) provides flexible networking models to support IP address management and traffic control. Networking allows additional functionalities such as load balancing, virtual private network (VPN) creation, and firewall configuration.

Together with core services, OpenStack also offers a set of juxtaposed shared services extending the core functionalities and contributing to their integration. Such a set of services comprehends:

- **Identity Service** (Keystone) provides a central directory of users, who are mapped to the OpenStack services they can access. The service acts as a common authentication system across the entire platform, including multiple forms of authentication such as username-password credentials, tokens, and AWS-style logins.
- **Image Service** (Glance) provides discovery, registration, and delivery services for disk and server images, giving the ability to copy, take snapshots of, and store server images. Stored images can be used as templates to rapidly instantiate multiple servers with the same characteristics, but they also ease software updating and modifications. The Image Service supports several image formats, among which are *OVF* used by VMWare, *VDI* from VirtualBox, and *VHD* used in (Hyper-V).
- **Telemetry Service** (Ceilometer) allows cloud operators to consult global or individual resource relative metrics.
- **Orchestration Service** (Heat) enables application developers to describe and automate the deployment of their cloud infrastructure thanks to a template language that allows users to specify both resources configurations and their workflow. It implements an orchestration engine to launch multiple composite cloud applications based on templates in the form of text files that can be treated like code. The template format defined by OpenStack, named **HOT**, is formally compliant with the *CloudFormation* Template defined by Amazon. An HOT file is an orchestration document that accurately describes all the elements required for the correct orchestration of multiple services, comprising: components, an abstract representation of the capabilities of a cloud service; resources, representing the artifacts of a deployment; definition of input and output parameters of each resource. Templates are described using YAML. The main advantage of HOT consists in the possibility to easily share it between multiple cloud providers, as templates contain vendor-independent specifications for launching a particular application in a target cloud environment.
- **Database Service** (Trove) quickly and easily provides relational database functionalities and capabilities, avoiding complex management issues.

The OpenStack community has also worked on reducing incompatibilities with Amazon EC2 and S3 APIs, in the attempt to further enhance the platform interoperability. The IBM commitment to the project may lead to interesting developments, given its support to the OASIS Topology and Orchestration Specification for Cloud Applications (TOSCA) TC, which is in sync with the OpenStack Heat project.

4.2.1 Access to OpenStack Services

OpenStack offers different possibilities to access and control its services. Surely the simplest and most appealing instrument to manage, the OpenStack platform is represented by the Web-based GUI also referred to as the **Horizon** Dashboard. The vista on resources changes according to the user's privileges. For system administrators who want to use scripts to control resources running on an OpenStack-enabled platform, command line interfaces written in Python are available. Each core service can be managed through a specific CLI. However, the functionalities offered by the compute dedicated interface (at the time of writing, the *novaclient*) are generally sufficient. Of course, RESTful APIs are also available, even if they frequently change according to the different OpenStack releases and some are still experimental. Exchanged messages are encoded in JSON.

4.2.2 Engagement with Case Study and Positioning with Respect to Use Case Scenarios and Features

The OpenStack offer includes only the IaaS level resources, so a direct and complete implementation of our Business Application test case would be impossible. Use case CSCC5 is thus not supported, unless we do not rely on third-party applications that can be deployed in the OpenStack infrastructure as virtual appliances. However, compute resources offered by OpenStack can be exploited to host applications like **Salesforce**. The principle of this public Software-as-a-Service is that customers do not have to purchase computing equipment, and can further reduce their expenses by sharing the cost of the infrastructure with other customers. Salesforce integration with OpenStack is made possible thanks to Nova API, which allows communication between the SaaS and IaaS resources. Through the Salesforce SaaS solution, it is then possible to interoperate with other cloud platforms such as Oracle, thus partially supporting CSCC2. In this scenario, OpenStack APIs can be used also to communicate with external applications through HTTP REST, for example to interface with Amazon EC2 and S3. PaaS solutions, such as Openshift, Cloud Foundry, and Bluemix, are able to directly interoperate with OpenStack. Use cases CSCC3 and CCSC4 are limited to the IaaS level: a customer can, for instance, substitute the computing capabilities of her privately managed infrastructure with Nova services which, in turn, can be easily connected to Amazon storage. Use case CCUC1 can be realized using OpenStack, since it offers APIs to migrate VMs from or to other platforms, provided they are packaged using a supported format. Figure 4.5 reports an example of implementation of the Business Intelligence application using the

Salesforce SaaS as a support. The positioning of the OpenStack solution over our
n-dimensional space is illustrated in Fig. 4.6.

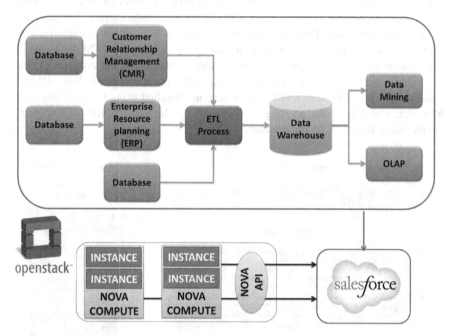

Fig. 4.5 Application design through OpenStack

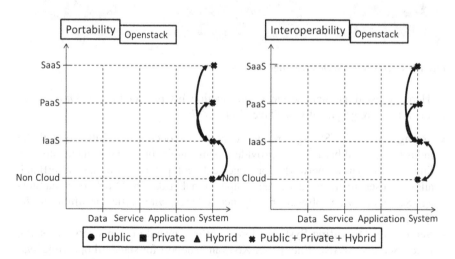

Fig. 4.6 OpenStack solution positioning

4.3 Oracle PaaS

The Oracle cloud platform [5] is a portfolio of products that can be used to build applications to publish as services on both private and public clouds. The platform is based on the **Oracle Grid** technologies, as well as on applications that include WebLogic Server, Coherence in-memory datagrid, and JRockit JVM. In terms of infrastructure, the platform is based on the Oracle IaaS offer that contains Oracle Solaris, Oracle Enterprise Linux, and Oracle VM for virtualization, Sun SPARC, and Storage. Both the IaaS and PaaS services are handled using Oracle Enterprise Manager, which provides an integrated system for the management of the entire development life cycle of applications.

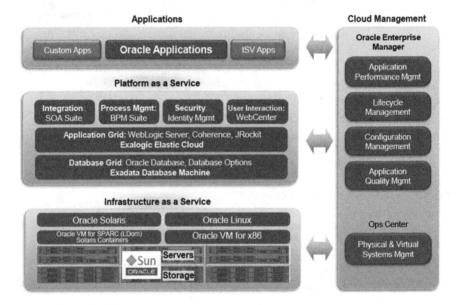

Fig. 4.7 Overview of Oracle PaaS offer

Here are listed the main services and functionality provided by the Oracle platform, as they are graphically displayed in Fig. 4.7.

- **Storage**: Sun Open Storage products combine the open-source software with the industry standard hardware to provide a platform for open and scalable storage. Sun provides virtual networks for large-scale computations through InfiniBand, allowing users to create very large computational grids. The Oracle Exadata Storage Servers provide software intelligence features, with particular affinity to the Oracle database.
- **Server and Operating Systems**: Oracle offers several of Sun's servers (Netra, Blade Servers, SPARC Enterprise, X64) and support for standard operating sys-

tems (Solaris, Linux, Windows), which provide a broad range of optimized physical infrastructures for virtualized and distributed nature of cloud applications.

- **Server Virtualization**: Oracle VM provides support for both the x86 and SPARC architecture, making possible the publication of applications in heterogeneous environments. Users can exploit Oracle VM to consolidate servers, release software quickly, recover quickly from system failures, and associate the ability to workloads.
- **VM Templates and Assemblies**: Oracle VM templates are virtual machine images containing preinstalled and configured enterprise software, which can be used to rapidly develop, package, and deploy applications. Oracle VM templates can speed up and simplify application deployments and help reduce the risk of errors in production, development, or test environments. Each VM template is essentially a software appliance because, like hardware appliances, they are prebuilt and very easy to deploy. The next level of this type of application packaging is the concept of VM Assemblies. While software appliances are useful, enterprise applications are not always self-contained, single-VM entities but are sometimes complex, multi-tier applications spanning multiple VMs. There may be multiple VMs in the web tier, other VMs in the middle tier, and other VMs in the database tier. There needs to be a way for these multi-VM applications to be packaged for easy deployment. Oracle virtual assembly builder is a tool that takes such a multi-tier, distributed application and packages it up into an assembly that can be reused in a way similar to the way appliances are used.
- **Database and Storage Grid**: Oracle database has offered grid computing capabilities since the release of Oracle Database 10g in 2003. Since then, Oracle has continued to enhance the grid capabilities of the database in the areas of clustering with Oracle Real Application Clusters (RAC), storage virtualization and manageability with Automatic Storage Management (ASM), and database performance with In-Memory Database Cache. When lighter-weight database services are needed, Oracle Berkeley DB and MySQL are also possible options that are actively developed and supported by Oracle.
- **Application Grid**: Similar to the grid architecture in Oracle database and storage, Oracle fusion middleware also supports robust grid functionality in the middle tier with a group of products called Oracle application grid. The key technologies that make up Oracle's application grid are Oracle WebLogic Server as the flagship application server; Oracle Coherence providing in-memory data grid services, JRockit JVM providing lightweight, lightning fast Java runtime environments; and transaction monitoring and management with Oracle Tuxedo.
- **SOA and Business Process Management**: Oracle SOA Suite provides a comprehensive yet easy-to-use basis for creating the reusable components at the heart of your PaaS private cloud. Rich drag-and-drop SOA component features in JDeveloper and the SCA designer enable rapid creation of components and subsequent composition of these components into applications. Oracle Service Bus provides a simple way to make components available to department application creators using the PaaS cloud. End-to-end instance tracking and Oracle Business Activity Monitoring provide a range of metrics visualizations supporting both the central IT

function charged with keeping the PaaS up and running and the departmental application owners concerned with business-level performance indicators. In addition to SOA components, many enterprises will want to include business process components managed within a unified BPM framework as part of their PaaS. Oracle BPEL Process Manager provides the federation capability to create BPEL process components out of new as well as legacy assets while also supporting the flexibility to enable multiple departments to incorporate PaaS-based BPEL components into their respective workflows.

- **User Interaction**: A centrally managed library of UI components can give department application owners a great head start in composing their solutions and gives the central IT function a desirable level of control over consistency across the enterprise's UIs. Oracle WebCenter Suite provides a number of portal and user interaction capabilities that are ideal for creating reusable UI components as part of a PaaS.
- **Identity Management**: Oracle Identity and Access Management Suite provides an ideal facility for managing access and security in a PaaS environment. Oracle Access Manager supports corporate directories and single sign-on. Oracle Entitlements Server provides centralized access control policies for a highly decentralized PaaS environment. Oracle Identity Manager is a best-in-class user provisioning and administration solution that automates the process of adding, updating, and deleting user accounts from applications and directories. Oracle Identity Federation provides a self-contained and flexible multi-protocol federation server that can be rapidly deployed with your existing identity and access management systems.

4.3.1 Engagement with Case Study and Positioning with Respect to Use Case Scenarios and Features

Oracle cloud allows the deployment of almost all of our Business Intelligence application's components, which find an equivalent at the IaaS, PaaS, or SaaS level. An example of how the test case, illustrated in Fig. 1.6, can be implemented using Oracle components is reported in Fig. 4.8. Oracle cloud does not include a data warehouse, which is however offered as a non-cloud solution, while CRM and ERP are available as SaaS. ETL, OLAP, and Big Data Analysis services are natively offered as hybrid solutions, making it possible to purchase them as cloud SaaS services that interact with the non-cloud data warehouse. A customer wanting to migrate a Business Application to the cloud could decide to retain her own data warehouse and use CRM, ERP, or ETL services offered by Oracle to work on her data. This kind of solution supports the CSCC4 use case. She could also decide to leverage the Oracle Data Warehousing offer, completely migrating all of her application components, thus realizing the CSCC5 use case. If we focus on CMR and ERP components, they can be mapped to SaaS services that can be chosen from the Oracle Marketplace: using services from different vendors is possible, since the Oracle platform takes care of their integration.

This enables interoperability between different providers' services (use case CSCC2) and it also enhances portability (use case CSCC1), since the customer can substitute one of the chosen services with another one from the same Marketplace. However, these possibilities are limited to the third-party services and applications which can be purchased from the Oracle Marketplace. Thanks to the Oracle support to different programming languages, first of all Java, each of the components of our Business Application could be developed exploiting the PaaS-level services. Another remarkable enhancement to interoperability is represented by OpenStack capabilities to be integrated into a broad range of Oracle products and cloud services. The positioning of the Oracle solution over our n-dimensional space is illustrated in Fig. 4.9.

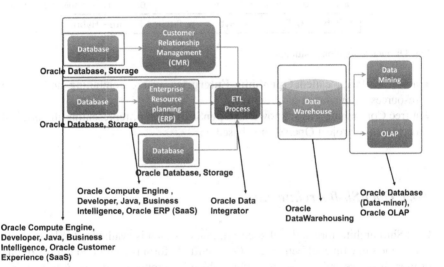

Fig. 4.8   Application design through Oracle

4.4 OpenShift

OpenShift [6] is a cloud computing platform provided by Red Hat, categorized as a platform service according to the NIST definition of cloud computing [7]. The main goal of the platform is to provide a cloud environment in which developers can easily and quickly design, develop, build, host, and scale applications using one or more of the programming languages and frameworks it supports. The user can choose among three different kinds of offers:

- a Public PaaS directly managed and supported by Red Hat, under the name of **OpenShift Online**;

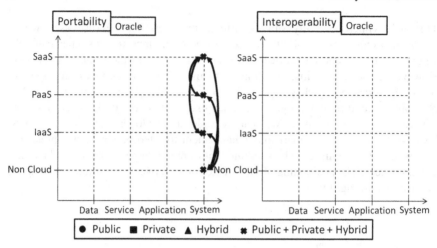

Fig. 4.9 Oracle solution positioning

- a Private PaaS, called **OpenShift Enterprise**, where computing and storage resources are managed on-premises;
- a free Community PaaS, known as **OpenShift Origin**, representing the original open-source project OpenShift is based on.

4.4.1 OpenShift Architecture

OpenShift architecture, a detailed description of which is available at [8], is simple and, at the same time, effective. The OpenShift platform is composed mainly of one or more **brokers** and a series of **nodes**, which are grouped into sets. A broker acts as a manager for a single set of nodes, and each set must be managed by a specific broker. Furthermore, a broker represents the only access point to the applications and frameworks running on each node: the user can only interact with the broker using the available OpenShift APIs, which allow the developer to manage every aspect of the developed applications, from hosting to scaling. Direct interaction with the node is not possible. A node is an instance of **Red Hat Enterprise Linux**, which is the foundation of OpenShift, and it represents the place where end-user applications reside. Every node is logically composed of **Gears**, which are automatically allocated and managed by the broker according to the computing and storage needs of the particular application hosted by the node, together with the policies set by the user through the provided API. A gear can be seen as a set of resources at the disposal of the node to execute an application residing on it. A platform administrator can set the dimensions (in terms of CPU speed, RAM, and disk space) and the number of gears a consumer can use for its applications, while the user decides to manually allocate them or let the platform automatically scale. A gear can support one or more **cartridges**, depending on the gear's dimensions and on the cartridge's requirements. A cartridge represents a particular feature or capability a developer can leverage when building its applications. Such features may include support to specific programming languages, databases, web, and application servers offered at the PaaS level and other services. Customers can add these features via both standard or custom cartridges, which can be designed and built by the user himself: this means that the developer can build his application in any available programming language, exploiting a framework of his own choice, and then he can host his application on a node of the OpenShift platform through the broker's API, which can be also used to select the needed cartridge.

4.4.2 Support to Portability

One major advantage of the architecture provided by OpenShift is represented by the possibility to run third-party applications on an OpenShift node just by selecting the correct cartridge and vice versa, thus strongly supporting portability. The limits imposed by the restricted number of available cartridges are easily overcome by the possibility to create custom ones. In particular, OpenShift addresses the possibility to seamlessly switch between development and operative environments, without the need to modify applications' code or introduce particular changes in a software's architecture.

4.4.3 Engagement with Case Study and Positioning with Respect to Use Case Scenarios and Features

A Business Intelligence application can be migrated to OpenShift with the use of a collection of cartridges. An example of how this can be realized is illustrated in Fig. 4.10. The CRM components can be replaced by the *JBoss Business Process Management* cartridge which is able to interoperate with external systems such as databases or ERP components. The database cartridge can be linked to several kinds of databases such as MySQL, MongoDB, or PostgreSQL. The ETL process can be replaced by the JBoss Data Virtualization component that enables the access to several sources of data such as Oracle DB, IBM DB2, Microsoft SQL Server, MySQL, Teradata, LDAP, SAP, Apache Hive, MongoDB, and others. The data warehouse system can be replaced by the *JBoss Data Grid* cartridge in combination with external resources for OLAP and data mining analysis. Natively, OpenShift does not offer a cartridge for an ERP system but it is possible to combine a *PostgreSQL* cartridge with a Python cartridge to run on the platform software like *Open ERP*. An alternative solution would be to integrate *JBoss Data Virtualization* with the Amazon solution *RedShift* to build the ERP system.

Considering all the different possible deployments of our reference application onto the OpenShift platform, we can say that scenarios CSCC1-5 are supported, even though only the PaaS and SaaS level services can be exploited. The positioning of the OpenShift solution over our n-dimensional space is illustrated in Fig. 4.11. The portability features of OpenShift are enabled through DeltaCloud API. Third-party software can be deployed in the OpenShift platform and are directly interface-able to OpenShift services. Through CloudFormation the interoperability between OpenShift and Amazon EC2 infrastructure is enabled. Some OpenShift services are able to interoperate with other PaaS services such as Bluemix database.

4.5 Microsoft Azure

Microsoft Azure [9] is an operative system considered to support cloud computing functionalities and capabilities. It enables the connection of several computing nodes, storages, and other physical/virtual resources to enhance resource sharing and reduce wastes. The Azure platform covers both the IaaS and PaaS layers. At the IaaS level, it manages storage and virtualized computation resources allowing users to completely control their infrastructure, providing a highly flexible environment. At the PaaS level, it offers a hosting environment for users' applications, which completely hides infrastructure details, thus relieving consumers from the burden of controlling and scaling the available resources.

4.5.1 Azure IaaS Level Services

The set of services provided at the IaaS level comprehends:

- **Virtual Machines**, the core of Microsoft Azure's IaaS solution, enabling infrastructure provisioning and the deployment of VMs on demand. Available VMs allow system administrators to deploy and configure both Linux and Windows Server images. Hybrid cloud solutions are fully supported thanks to the possibility to connect VMs running on the Azure platform with one running within an in-premise environment. Also, VM migration is possible, provided the VHD format is used. Connections with other Microsoft services such as Microsoft SQL Server and SharePoint Server, and external databases and storages such as Oracle, MySQL, Redis, and MongoDB are also possible.
- **Storage, Backup, and Recovery** include a plethora of services dedicated to data storage, security, and replication. **Azure Storage** is a scalable, elastic, and ubiquitous storage for massive data, available in different flavors:

  - **Blob storage** is able to store any type of text or binary data, such as a document, media file, or application installer.

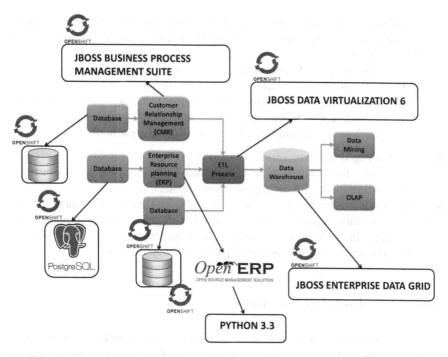

Fig. 4.10 Application design through OpenShift

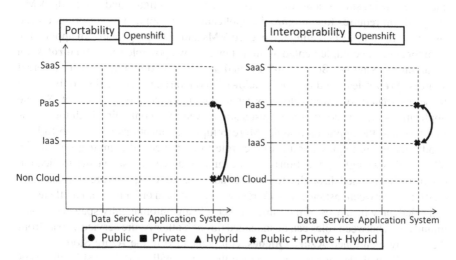

Fig. 4.11 OpenShift solution positioning

- **Table storage** is a NoSQL key-value data store that allows for rapid development and fast access to large quantities of data. The storage comes with a well-defined tabular structure.
- **Queue storage** is used for reliable communications based on message exchanges between components of cloud services.
- **File storage** offers shared storage capabilities for legacy applications using the standard **SMB 2.1** protocol. This shared storage can be used by both Azure virtual machines and cloud services and on-premise applications.

- **Azure Backup** is a backup service offered by Azure that ensures data protection and encryption. It collaborates with the **Site Recovery** service to guarantee disaster recovery and resiliency for both applications and data.
- **Big Compute** is a service created to deliver high performance and scalability, typical of on-premise supercomputers, through the cloud. Highly performant compute instance in the Azure environment can access remote direct access memory (RDMA) technologies to be used with parallel MPI applications.

4.5.2 Azure PaaS-Level Services

The core of the Azure PaaS offer is represented by **cloud services**, enabling customers to easily deploy applications on the Azure platform which, in turn, manages their scalability, elasticity, and replication and ensures their availability. Cloud services rely on VMs for the execution of applications, but it differs from the virtual machine service since clients do not need to actually configure and launch the VMs: the platform requires to configure the application with some basic information and then it takes charge of creating the needed VMs and managing them. An application component can be implemented as an instance of two possible roles: **Web roles** run a variant of Windows Server with IIS and are able to communicate with external sources; **Web roles** run the same Windows Server variant without IIS, so they need to be associated to a web role in order to communicate through the net. Simple applications may be composed of just a single web role, containing both the access interface and the application logic. More complex applications use a Web Role to handle incoming requests and a worker role to do the actual computation. All the VMs used by a single application run in the same cloud service, so that the application can be fully accessed through one single IP address, while requests are automatically distributed between instances of different worker roles, in order to avoid bottlenecks. If needed, new instances of the same worker role are automatically spawned. Communications between roles can be managed through other Azure services: apart from the already mentioned **Azure Queues**, comprehended in the Azure Store services at the IaaS level, developers can leverage the functionalities delivered by the PaaS **Service Bus**. Service Bus delivers the basic queue functionalities, but also provides additional functionalities, like the possibility to mark messages, in order to let consuming worker roles decide how to act on them. Also, roles can subscribe to the bus so that they need not continuously poll the queue in order to receive a message.

4.5.3 Engagement with Case Study and Positioning with Respect to Use Case Scenarios and Features

An example of how our test application can be deployed in Windows Azure is reported in Fig. 4.12. In particular, it is possible to leverage a composition of ClearDB (SQL Database), SugarCRM (compliant with IBM DB2), Sage 300 ERP (offering a lot of import features from external applications), HdInsight (ELT), Birst Analytics HdInsight (Data warehouse), and Birst Analytics (Data Mining and OLAP). The positioning of the Azure solution over our n-dimensional space is illustrated in Fig. 4.13. Exploiting services mostly at the SaaS level, the portability is guaranteed by the use of third-party software, mostly compliant with other cloud providers. If we consider the possibility to develop the application at the PaaS level utilizing a common programming language supported by the platform (such as Java), the portability features extend to all the cloud platforms that support the same languages, assuming to change the code portion that belongs to Windows Azure-specific services. Concerning data portability, mechanisms enabling the transfer of data between azure storage services exist, as well as export/import features for external SQL-based databases. For what concerns interoperability, through the offered REST API it is possible to interoperate between different components deployed over private or public clouds at the IaaS and PaaS levels. Due to the above-mentioned characteristics, the Azure solution addresses the use case scenario CSCC S2 for what concerns interoperability and some aspects of the use case scenario CSCC S1 for what concerns portability.

4.6 Google Cloud Platform

Google has been among the first providers to offer a set of PaaS services to consumers. Because of this, they have established as a "de-facto standard" in the PaaS scenario, much like how the already cited Amazon (Sect. 4.1) has done for IaaS. Recently, both Google Cloud and Microsoft Azure have added IaaS services to their portfolio. In the following subsections, we will introduce the services included in the Google Cloud Platform [10] offer, focusing on the interoperability and portability features they expose.

4.6.1 Google Compute Engine

Google Compute Engine represents the IaaS offer provided by Google, enabling users to create VMs of various dimensions and administer them through the **Google Console**, a well-documented API or via command line tools. The service, previously only available through invitation or after a direct contact with the Google sales team, is now publicly accessible for customers signing up for a Google account and

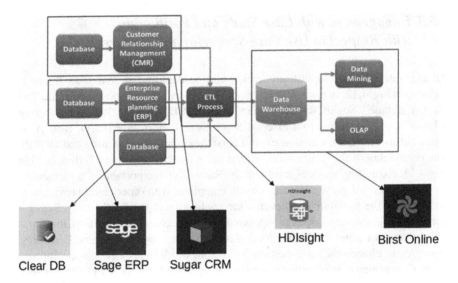

Fig. 4.12 Application design through Azure

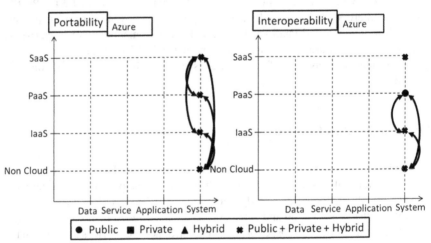

Fig. 4.13 Azure solution positioning

setting their billing preferences. As time passes, features and services offered by Compute Engine widen: for example, in previous versions the available operative systems available only included **Google Compute Engine Linux** (GCEL), which is a Debian fork optimized for the use on Google Cloud and CentOS. As of now different OSs are available (GCEL has been deprecated), including CentOS and Debian as standard Linux images, while Red Hat Enterprise Linux (RHEL) and Microsoft Windows Server 2008 R2 are offered as a part of the premier operating system images available for an additional fee. KVM is the hypervisor used to run VMs, which need a **Persistent Disk** to boot. The number of virtual CPUs and the amount of memory supported by the VM depend on the machine type selected. Prices vary according to the selected resources. When a virtual machine instance is launched, an **Instance** resource is created that uses other resources, such as:

- **Disk**, representing a persistent memory completely independent of any connected VM.
- **Snaphot**, a backup of the data contained in a persistent disk, which can be used to restore lost data or to migrate them across different disks.
- **Network** connects instances with each other and with the outside world, by defining a range of available addresses and gateways.
- **Address** represents the IP of an instance. An ephemeral address is automatically assigned to each launched instance and it is released when the instance is terminated. However, static IPs can also be reserved to VMs.
- **Firewall** contains rules that filter connections and communications between instances.
- **Route** represents a routing table, working similarly to a physical router, managing traffic and data paths within a network.

Analogous to AWS, also Google Cloud provides a location-based distribution of resources, which are generally associated to **Regions**, referencing to geographical areas where Google facilities reside, and **Zones**, representing isolated locations within a Region. The Google Compute Engine currently supports different third-party services and software. Interoperable services include:

- **CloudAMPQ**, a managed service that offers hosted RabbitMQ as a service in the cloud.
- **MongoLab**, which manages the open-source NoSQL database known as MongoDB.
- **Redis Labs**, offering managed Redis databases in the cloud. Redis is an open-source advanced key-value cache and storage software.

Supported software mostly fall under two categories:

- **NoSQL databases**, including Apache Cassandra, DataStax Enterprise, and MongoDB.
- **Configuration management**, including Ansible, Chef, Puppet, and SaltStack.

4.6.2 Google Cloud Storage

Google Cloud Storage is an IaaS Web service, comparable with the Amazon S3 online storage. It can be accessed online through a RESTful interface for storing and retrieving data on Google's infrastructure. Other access means are by a web-based interface, a command line tool, and several language libraries. In order to manage controlled accesses to stored data, Google Storage exploits access control lists (ACLs), consisting of a set of entries granting or limiting certain privileges (for example, read and write) to users when accessing data. The storage service is fully compatible with other IaaS and PaaS offers from Google, and it offers interoperability with Amazon S3 and Eucalyptus.

4.6.3 Google App Engine

Google App Engine (also GAE or simply App Engine) represents the core product of the Google cloud offer and probably the most popular and interesting service from Google. It is a PaaS product, consisting of a developing platform for hosting applications written in one of the supported programming languages, such as Python, Java (and other JVM languages such as Groovy, JRuby, Scala, Clojure), Go, and PHP. Support for the last two languages is still in an experimental phase. The App Engine service allows programmers to develop, run, and test their applications in secured sandboxes, which automatically manage application deployment and resource scaling, according to the current compute power and storage volume requested. The use of resources is free until a fixed quota is reached: after that, additional fees are applied. Given its support to Python and Java, the App Engine offer includes support to a set of notable frameworks, even if workarounds are sometimes needed for Java. Python web frameworks that run on Google App Engine include Django, CherryPy, Pyramid, Flask, web2py, and webapp2 but, in general, any Python framework that supports the WSGI using the CGI adapter can be used to create an application. As regards Java, App Engine fully supports the open-source Jetty Web Server and its related technologies (such as JSP). The Java Persistence API (JPA) and Java Data Objects (JDO) are both provided in order to access, read, and write to datastores. While the Spring and Struts 1 framework are both supported, JavaServer Faces and Struts 2 need workarounds to correctly run on App Engine. When developing a Java application, programmers can leverage the Eclipse IDE: this is possible through the officially provided plug-in, which extends the IDE enabling users to access tailored project templates and test code on a simulation environment reproducing the target platform's behavior. Different storage solutions, including the already mentioned Google Storage, are available to developers:

- **Google Cloud SQL** offers access to a MySQL database, which guarantees to avoid data lock-in thanks to the standard connections and tools (mysqldump, MySQL Wire Protocol, and JDBC) which can be used to migrate onto or off the Google Cloud Platform. Data are automatically replicated across multiple regions to ensure continuous availability and failure resiliency.
- **Google Datastore** is a schemaless storage, accessible through a RESTful API allowing the execution of queries. The datastore holds entities that are data objects

with associated properties. Entities are categorized (for query support) and are identified by keys. The datastore supports transactions containing several data operations: in case of failure of a single operation, the entire transaction is rolled back.

- A **Blobstore API** allows developers to work with large files, called blobs, the dimension of which exceeds that allowed by the datastore service. Blobs are associated to a key, which is used to retrieve them.

Despite the possibility to migrate data from Google stores and databases and the support to Python and Java frameworks, programmers have often expressed their fear of being locked-in by Google technologies because of the proprietary/closed APIs offered by the App Engine, especially for accessing the datastore service. Different projects have arisen to solve possible portability problems, being AppScale [11] and TyphoonAE [12], two of the most mature open source efforts.

4.6.4 Google BigQuery

Google BigQuery is a Web service, considered an IaaS, which lets consumers perform interactive analysis of massive datasets. Access to both BigQuery projects and datasets is possible through the Google APIs console, browsers, or Command Line tools. One of the main features of the service is the possibility to load data directly from the Google cloud storage, ensuring the fast migration of very large datasets. However, data can also be loaded from external sources, provided they are formatted using JSON or CVS. Since the analysis of the huge datasets typically involved in big data applications could require long execution times, queries written in BigQuery's SQL dialect can be run asynchronously and customers can access the system to check their advancement status.

4.6.5 Engagement with Case Study and Positioning with Respect to Use Case Scenarios and Features

The Business Intelligence application we are using as a case study would be developed on Google platform, using Google Cloud Platform, Google Cloud Storage, and Google BigQuery for what concerns the data warehouse, ETL and OLAP components. The ERP and CRM processes would be implemented using third-party software in the Google Compute Engine or using the SaaS such as BaseCRR or Simple ERP and CRM. A sample design of this application on the Google platform is illustrated in Fig. 4.14, thus fully implementing use case CSCC5. Since the Google PaaS platform offers support to different programming languages, Java in particular, porting of application to and from it should be possible as long as the exploited programming language is supported (use case CSCC1). Google has recently released a preview of its **Google Online Cloud Import** service, allowing import of data from third-party databases, Amazon storage included. For what concerns interoperability, the use of Google services is strictly related to the use of the Google Cloud Platform, so interoperability with other providers' services is not always guaranteed. SaaS products provided by Google generally offer openly accessible APIs to interact with. At the IaaS level, Google Cloud Storage guarantees interoperability with Amazon Simple Storage Service (Amazon S3) and Eucalyptus Systems. A user can easily decide to move only part of her application to the Google platform (use case CSCC4): for example, databases and data warehouse can be substituted by Google Storage services. The positioning of the Google solution over our n-dimensional space is illustrated in Fig. 4.15.

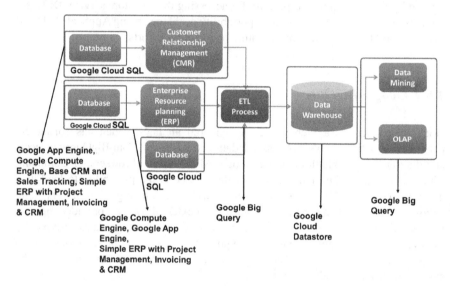

Fig. 4.14 Application design through Google

4.7 Bluemix

Bluemix [13] is IBM's cloud-based public PaaS environment for building, managing, and deploying various kinds of applications, either web or mobile. It delivers a set of prebuilt services and hosting infrastructures for application deployment, as well as business logics, development back-ends, and monitoring capabilities. Bluemix is built in using the CloudFoundry [14] open-source technology, extending community-related projects with IBM's set of services. Being a PaaS platform, application development, delivery, management, and availability is provided by abstracting and completely hiding the traditional complexities associated with the hosting- and management-based applications on a cloud infrastructure. It is thus possible to focus on the application development, without caring about the infrastructure management required to host the services exploited by the application. After applications have been deployed in the Bluemix platform, it is possible to quickly scale up or down according to the applications' workload fluctuations.

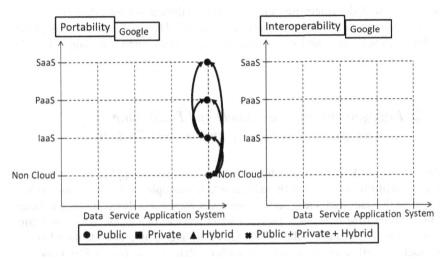

Fig. 4.15 Google solution positioning

4.7.1 Overview of the Offered Services

Although a very young project, which was launched in beta version at the beginning of 2014 and just recently released (at the time of writing), Bluemix already offers a wide catalog of services, classified in the following categories:

- **Runtimes** include build-packs created by IBM itself, such as Liberty for Java and Node.js; and community build-packs such as Ruby on Rails and Ruby Sinatra.
- **Web and Application** comprehends a wide range of services helping in the application development. These include, but are not limited to, cache services, queues management, log analysis, and integration (DataCache, Session Cache, ElasticMQ, Rules, Single-Sign-On, Travel Boundary, Validate Address, Reverse Geocoding, Geocoding, Redis, RabbitMQ, RapidApps, Cloud Integration, CloudAMQP, Redis Labs, SendGrid, Application Auto-Scaling, Log Analysis, Twilio).
- **Mobile** services are meant for mobile application development, including application security, quality assurance, and notification services (Push, Internet of Things Cloud, Mobile Data, Mobile Application Security, Mobile Quality Assurance, Square).

- **Data Management** includes different kinds of databases (SQL Database, Cloudant NoSQL Database, ClearDB, ElephantSQL, MongoDB, PostgreSQL, MySQL).
- **Big Data** with analysis services such as Analytics Warehouse, Analytics for Hadoop, and Time Series Database.
- **DevOps** includes enterprise capabilities for continuous software delivery and life-cycle management (Monitoring and Analytics, Git Hosting, Web IDE, Continuous Integration, Continuous Delivery Pipeline, Agile Planning and Tracking, BlazeMeter, Load Impact).

4.7.2 Engagement with Case Study and Positioning with Respect to Use Case Scenarios and Features

An attempt to develop, using the Bluemix platform, the application example of Fig. 1.6 is illustrated in Fig. 4.16. In Bluemix, the example application would be built by using the data warehouse service *Analytics Warehouse* that includes a database DB2 and the IBM *Infosphere Data Architect* services that enable database porting to the cloud. DB2 can be linked to OpenStack through a plug-in or deployed in EC2 through an AMI or migrated to/from Oracle database via the *Oracle SQL Developer* or *IBM Data Movement Tool*. Business services such as ERP, CRM, and ETL are not directly offered by Bluemix but alternative services are able to interoperate with the Bluemix platform through the *Embeddable Reporting* service. The positioning of the Bluemix solution over our n-dimensional space is illustrated in Fig. 4.17. Bluemix is a public platform, so only portability toward or from public cloud is relevant. Bluemix allows portability from the PaaS to SaaS, since it is possible to rely on some software applications without the need for a database configuration, which may be provided as part of the service itself. Intra- and interportability from the SaaS to PaaS is supported as well since it is always possible to add a database service to the application. Bluemix takes advantage of Cloud Foundry portability solutions: in Cloud Foundry, the components of a PaaS offering that applications depend on (e.g., runtimes, messaging, data access) are built using open development frameworks and technologies (Java, Ruby, Node.js, MongoDB, MySQL, PostgreSQL, RabbitMQ, Redis). Portability from a non-cloud application to a PaaS environment is possible in Bluemix since IBM on-premise services can be easily transferred to the corresponding cloud service, when an equivalent exists. Migration in a SaaS environment is possible also since, for example, business intelligence reports derived from on-premise services can be integrated into a Bluemix application through the Embedded Reporting cloud service. In addition, the Bluemix analytics warehouse service includes the IBM infosphere data architect tool for the quick migration of DBs into the cloud. Interoperability with different providers is possible thanks to the possibilities offered by the DB2 service: in fact, it can be made available in an OpenStack environment and it is also offered as an Amazon AMI, allowing deployment in EC2 virtual machines. Thus, interoperability between the PaaS and IaaS is possible, both using public or private infrastructure. Interoperability with the SaaS services is also possible since each Business Intelligence ETL tool (third-party or not) compatible with DB2 is also compatible with the Bluemix analytics warehouse service. Some Cognos SaaS services (also offered as on-premise services) may also rely on different database systems such as DB2, Microsoft SQL Server, and database Oracle. Note, however, that the SQL database service running on Bluemix is accessible only to applications running in Bluemix itself. Bluemix can address some aspects that concerns the use case scenario CSCC S1. In particular, portability from PaaS to PaaS is possible thanks to the database possibilities since migration between different SQL dialects is possible through the use of proper conversion tools. Migration toward Oracle databases, for example, is possible through the Oracle SQL Developer tool, whereas migration toward IBM DB2 is allowed through the IDM data movement tool. Bluemix can also address some aspects that concern the use case scenario CSCC S5. In particular, the Bluemix analytics warehouse service includes the IBM infosphere data architect tool for quick migration of DBs into the cloud. Also, some interoperability aspects that concern the use case scenario CSCC S2 can be addressed by Bluemix. In particular, interoperability with different providers is possible thanks to the possibilities offered by the DB2 service. In fact, it can be made available in an OpenStack environment, through the use of some tools and proper settings; and it is also offered as an Amazon AMI, allowing deployment in EC2 virtual machines. Thus, interoperability between the PaaS and IaaS is possible, both using public and private infrastructures. Interoperability with the SaaS services is also possible since each Business Intelligence ETL tool (third-party or not) compatible with DB2 is also compatible with the Bluemix Analytics Warehouse service. For what concerns hybrid scenarios, such as the case described in CSCC S4, the Bluemix cloud integration service can be used for the deployment of applications in a hybrid environment. It uses secure connectors to talk securely to applications running behind a firewall. Once a tunnel is established, the Bluemix app can use the cloud integration service to access data from the server behind the firewall. Cloud integration for Bluemix enables you to integrate cloud and on-premise applications. The cloud code leverages the cloud integration service to interact with the backend databases such as DB2, Oracle, and SAP to create database APIs.

4.8 ElasticBox

The approach taken by **ElasticBox** [15] is to allow enterprises to create and manage applications using "Boxes," representing encapsulated, individually configured application stacks, meant to be reusable, easily transferable from platform to platform, and available as a service. Boxes contain application resources and components such as databases, language runtimes, web servers, and middleware. A box can be created by the user according to her personal needs or built by choosing

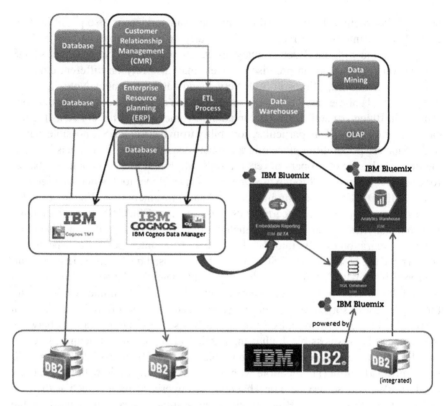

Fig. 4.16 Application design through Bluemix

components from a public catalog, comprehending boxes for Hadoop, Ruby, MongoDB, Python, and others. The creation of a multi-tier application architecture is possible through the combination ("stacking") of boxes.

The most appealing feature provided by ElasticBox consists in the possibility to develop, test, and run applications on different cloud platforms, with the only limitation represented by the actually supported platforms, currently comprehending a wide range of solutions including Microsoft Azure, OpenStack, Rackspace, VMware, Amazon Web Services, and Google Compute Engine. The selection of the target platform happens after the user has chosen the box to deploy: ElasticBox automatically configures the selected box in order to run on the specific platform (provided the user has the right to access it). Boxes also ease updating of applications: the user only needs to set Box policies to decide how and when updates will be spread and which applications will be affected. Once boxes have been deployed, there are two options to interact with them in order to manage and tweak running applications. The first possibility is to leverage the REST API offered by ElasticBox, which provides basic functionalities to work with running instances (stop, relaunch, delete), but also to interact with single boxes. However, in order to gain access to more advanced options, using the provided **Lifecycle Editor** tool is mandatory. The editor

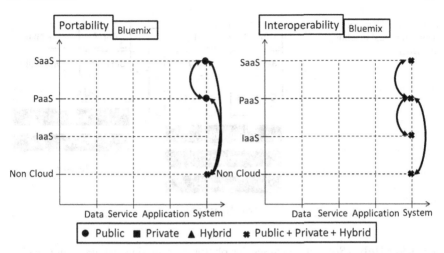

Fig. 4.17 Bluemix solution positioning

offers a single, unified interface to test and refine box configurations in live deployments. Changes made to box instances are immediately applied and can be monitored through logs. The possible interactions with Box instances include, together with the basic operations available through the API, the creation of event-based scripts and modifications of system variables.

4.9 Docker

Docker [16] presents itself as an open platform that both developers and system administrators can leverage to build, package, and deploy applications, with no limitations regarding the target machine, programming language, or operative system. Deploying on a cloud platform is only one of the possibilities offered by Docker, since applications can be packaged in **containers** that can be deployed anywhere. Each application can run securely isolated in a container, with multiple containers potentially running simultaneously on the same host without the intervention of a hypervisor. Figure 4.18 shows the difference existing between an architecture using a hypervisor to run a virtual machine (on the left) and Docker (on the right). While VMs need hypervisors hosted on a local operative system to run one or more guest OSs, where applications and relative library live, the Docker solution is directly installed on the local OS and does not need any further software to make containers work. In particular, the Docker consists of two main elements:

- **Docker Engine**, representing a portable, lightweight runtime and packaging tool, operating as a container virtualization platform.

Fig. 4.18 Difference between virtual machines and Docker

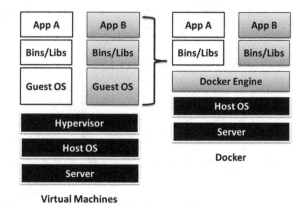

Virtual Machines

- **Docker Hub**, a Software-as-a-Service platform for sharing and managing Docker containers and applications and automating workflows.

Docker's architecture is based on the client–server paradigm: a **Docker Daemon** is responsible for building, running, and distributing Docker containers, while a **Docker Client** communicates with it to issue commands through sockets or REST APIs. Users cannot interact with the daemon directly, but they have to leverage the functionalities exposed by the client. Client and daemon can run on the same host or the latter can be contacted remotely indifferently from the users' point of view.

4.9.1 Internal Components

In order to understand how Docker works, it is necessary to introduce the internal elements used by the platform to deliver and run containers and applications within. In particular, Docker defines three main internal components:

- **Docker images** are read-only templates from which Docker containers are launched. An image is composed of layers, combined together through the Union File System (UnionFS) allowing files and directories of separate file systems to be transparently overlaid. When a Docker image is modified, for example by updating an application, a new layer is added or an existing one is updated: in this way Docker avoids to rebuild and redistribute the entire image. Every image is built on an existing base image, containing a default operative system, which can be selected from the Docker Hub or uploaded by the user directly.
- **Docker registry** represents a storage for Docker images. Such storage can be either public or private. Public registries can be searched from the client and images stored in them can be freely downloaded and used. Private registries are excluded from researches and only authorized users can pull images from them. Docker Hub provides both public and private registry support.

- **Containers** are composed of an operating system, some optional user-added files, and meta-data. Each container is built from an image, which tells Docker what the container holds, eventual processes to run during start-up of the container, and configuration data. Since a Docker image is read-only, when instantiating a container from it, the platform automatically adds a read-write layer on top (using UnionFS) in which the application can run. Docker extends a common container format called Linux Containers (LXC), with a high-level API providing lightweight virtualization that runs processes in isolation, in order to provide an internal container format. However, LXC is supported and other formats will be integrated in the future.

4.10 Cloudify

Cloudify [17] can be defined as a Cloud Application Orchestrator which aims at organizing the workflow of any kind of application, from the simplest to the most complex ones. All processes needing to run an application over a cloud platform can be automatized through Cloudify: this comprehends environment setup, application installation and update, infrastructure management, auto-scaling, and failure recovery. The creation of the whole cloud infrastructure needed to run a specific application is thus completely covered by Cloudify capabilities, starting from provisioning of compute resources to network and block storage configurations. Cloudify also takes care of the deployment of the applications to the cloud, with particular attention to the OpenStack platform, for which it provides native integration. In particular, it offers full support for OpenStack API and TOSCA templates (see Sect. 5.1) giving the possibility to test applications in a safe environment before their final deployment to OpenStack. Thanks to the standard description of cloud components orchestration provided by TOSCA, Cloudify surely reduces vendor lock-in issues and improves interoperability between existing infrastructures and cloud platforms. **Cloudify's architecture**, moving to release 3.0 at the time of writing, is composed of three main elements, which interact to deliver the platform capabilities. The **CLI client** is a Python executable that can run on Windows, Linux, and Mac OS, representing the main tool at the users' disposal to manage the deployed applications and the Cloudify manager itself. The two main functions provided by the CLI are indeed represented by:

- Manager Bootstrapping—Specific components belonging to the CLI, known as "Providers", are responsible for using a particular set of IaaS API to set up the networking, security, and VM environment required by the Cloudify manager and install the manager itself. The user is given the option to use the CLI to perform such a task or to exploit another tool of her choice.
- Managing Applications—The Cloudify manager interface cannot be accessed directly: the CLI acts as a REST client, providing users with a set of functions to deploy, manage, and monitor applications through logs.

The **Manager** is a stateful orchestrator taking care of the deployment and management of applications, whose workflow is fully described in orchestration plans known as **blueprints**. The manager interacts with **agents** defined by the Cloudify platform, in order to run processes defined in the workflow of an application. It is a complex element of the Cloudify platform, which contains a set of interacting components:

- Proxy and file server: Cloudify uses Ngnix as its reverse proxy and file server.
- REST API are used to control Cloudify, offering cloud orchestration and management functions. The API can be used through the provided CLI client, but customers can also create their own REST API.
- A Web GUI can be used as an alternative to the API, but it also adds additional functionalities and views on the system. For example, it offers graphical screens of the available blueprints, of the system and topology, or of the performances of the running applications.
- The workflow engine is used to manage applications' behavior through workflow descriptions. Timing and orchestration of tasks, responsible for creating and managing applications, are taken care of by this component. In order to accomplish the full orchestration of applications, the engine first interacts with blueprints to get the necessary information and then issues tasks to a broker, which is based on the Celery tasks broker.
- The policy engine enforces custom policies to make runtime decisions on availability, SLA, scalability, and so on.

The execution of the manager's command is enforced through **agents** using a set of plug-ins. Two categories of agents exist in the Cloudify architecture:

- Manager side agents are responsible for the application deployment, since they handle IaaS-related tasks such as the creation of virtual machines and networks. A manager side agent is responsible for the deployment of a single application deployment.
- Application side agents are optionally located on application VMs: the user decides if a VM will have an agent installed on it by stating this fact in the application blueprint. The application-side agents are installed on the VM as part of the VM creation task by the manager side agent. Plug-in installation and operation execution, such as task for configuration and deployment of application modules, are the responsibility of application side agents.

References

1. Amazon web services. http://aws.amazon.com/
2. Amazon web services—regions and availability zones. http://docs.aws.amazon.com/ AWSEC2/latest/UserGuide/using-regions-availability-zones.html
3. Cloudformation templates. http://aws.amazon.com/cloudformation/aws-cloudformation-templates/
4. Openstack services. http://www.openstack.org/software

5. Demarest, G., Wang, R.: Oracle Cloud Computing. Oracle White Paper p. 22 (2010)
6. Openshift by Red Hat. https://www.openshift.com/
7. Mell, P., Grance, T.: The NIST Definition of Cloud Computing. Recommendations of the National Institute of Standards and Technology. Computer Security Division, NIST, Gaithersburg, MD (2011)
8. Openshift slideshare channel. http://www.slideshare.net/openshift
9. Windows Azure web-site. https://azure.microsoft.com/
10. Google Cloud Platform. https://cloud.google.com/
11. Appscale: freedom for your applications. http://www.appscale.com/
12. Typhoon app engine. https://code.google.com/p/typhoonae/
13. Bluemix web-site. https://ace.ng.bluemix.net/
14. Cloudfoundry foundation. http://cloudfoundry.org/index.html
15. Elasticbox. https://www.elasticbox.com/
16. Docker—build, ship and run any app, anywhere. https://www.docker.com/
17. Cloudify—cloud orchestration and automation made easy. http://getcloudify.org/

Chapter 5
Research Initiatives and Emerging Standards

5.1 European Commission Initiatives

Establishing a coherent framework and conditions for cloud computing services in Europe, creating the world's largest cloud-enabled ICT market, is one of the objectives of the European Commission's Digital Agenda for Europe (DAE) [1]. It aims to reboot Europe's economy and help Europe's citizens and businesses to get the most out of digital technologies. It is the first of seven flagship initiatives under Europe 2020, the EU's strategy to deliver smart sustainable and inclusive growth. Among the actions determined by the DAE, the European Cloud Computing Strategy [2] is developing the cloud computing vision for Europe and the future research and policy directions. As part of this strategy, the European Commission has engaged a group of experts who have analyzed the current technological progress in the domain of cloud computing, have identified the major gaps and necessities for future research and development in cloud technologies [3], and have defined a roadmap for advanced cloud technologies under the Horizon 2020 research framework [4], including portability and interoperability.

5.2 Topology and Orchestration Specification for Cloud Applications

Topology and Orchestration Specification for Cloud Applications (TOSCA) [5] is an OASIS standard language used to describe both a topology of cloud-based Web services, consisting of their components, relationships, and the processes that manage them, and orchestration of such services, which is their complex behavior in relation to other described services. The combination of topology and orchestration, in what the standard defines as **service template**, accurately describes all the essential elements needed by each service to provide its functionalities, in order to

© The Author(s) 2015
B. Di Martino et al., *Cloud Portability and Interoperability*,
SpringerBriefs in Computer Science, DOI 10.1007/978-3-319-13701-8_5

ease deployments in different environments and to enable interoperability. Also, the management of services throughout their complete life cycle (deploying, scaling, updating, monitoring, ...) when applications using them are ported to different cloud platforms is also supported. In synthesis, TOSCA's purpose is to enhance portability and interoperability of cloud applications, and related IT services, by defining an interoperable description of cloud services, of the relationships existing among components of these services, and of their operational behavior, which in a way is independent of the cloud provider offering the services and of the technologies involved. In particular, the TOSCA technical committee has used as a starting point the "Topology and Orchestration Specification for Cloud Applications" document submitted by a number of cloud vendors (among which IBM and Red Hat are remarkable contributors) through a process of revision and extension of the existing XML Schema. The description of orchestration exploits existing workflow languages, with particular focus on BPEL [6]. The scope of the project also involves the ability to use virtual images, application artifacts, and off-the-shelf components as deployment artifacts for parts of a service template. The deliverables of the project include an extended version of the existing topology proposal, together with a set of sample cloud service templates to use for testing the conformance of individual TOSCA implementations as well as interoperability between different implementations. It is important to remark that TOSCA mainly focuses on the description of services and of their relationships, not on the IaaS infrastructure. Thus, it could be used to define cloud components and services at different layers. To specifically manage the infrastructure, other standards such as CIMI are more suitable: a provider could easily manage the cloud infrastructure required by a service described using TOSCA with CIMI (see Sect. 5.3).

5.2.1 TOSCA Architecture and Components

TOSCA defines a metamodel to describe IT services, which includes both its structure and how to manage it. A **topology template** (also referred to as the "topology model" of a service) focuses specifically on the structure. **Plans** describe the services orchestration, including the processes used to manage their entire life cycle from start to termination. The combination of a topology template and plans constitutes a **service template**.

A **topology** template can be represented as a directed graph, not necessarily connected, in which nodes are defined by **node templates** and arcs correspond to **relationship templates**. Each node template represents an instance of a **node type**, which defines all the properties (**node type properties**) and operations (**interfaces**) a certain component of a service owns and exposes to users for manipulation. In order to enhance reuse of components, node types are defined separately: a node template simply references a node type and adds constraints to its use, like the maximum number of times the described component can occur in a topology. Figure 5.1 shows the relations existing between templates and types.

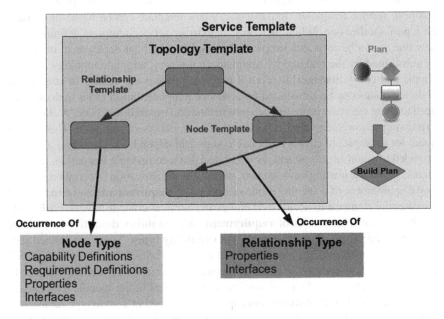

Fig. 5.1 TOSCA topology template

Relationship templates play a similar role as node templates, but they specify the occurrences of relationships existing between nodes in a topology template instead of nodes themselves. Each relationship template refers to a relationship type, defining a set of properties and semantics of the relation. **Plans** included in a service template define the management processes needed to start, terminate, and operate a service, generally described as a workflow of one or more steps. In order to enhance portability and interoperability, no new standard has been proposed to represent such workflow: instead, well-known technologies such as BPEL and BPMN [7] are used. However, TOSCA allows for any workflow definition language to be used. In order to realize the deployment of a service, TOSCA allows to associate artifacts to the defined templates. An artifact can be an executable, a configuration or data file, or even a library needed by another artifact, which are necessary for the concrete realization of a service. Attached to an artifact, it is possible to find meta-data describing information needed to correctly process it inside the execution environment. Two typologies of artifacts are available in TOSCA:

- Deployment artifacts describe the operations to execute when instancing a node.
- Implementation artifacts represent the executable of an operation of a node type.

5.2.2 Composition of Service Templates

One interesting feature provided by TOSCA is represented by the possibility to compose service templates. In particular, a new template can be based on or built on top of another one. When including an existing service template definition in a new one, it can be seen as a simple node template which, at deployment time, will be substituted by the real service template. In this way, template topology remains simple and easily manageable, even if multiple services are nested. In general, all node templates can be substituted by a service template, but in order to do so they need to share the same properties, requirements, and capabilities. TOSCA allows for expressing requirements and capabilities of components of a service. This can be done, for example, to express that one component depends on (requires) a feature provided by another component, or to express that a component has certain requirements against the hosting environment, such as for the allocation of certain resources or the enablement of a specific mode of operation. **Requirements** and **capabilities** are essential elements to decide over Nodes and Services compatibilities. Each Node can be annotated with a set of **requirement** and **capability definitions** which, in turn, are occurrences of **requirement** and **capability types**. Types are defined separately and can be reused in the context of different nodes. Node templates, which are occurrences of node types defining requirement or capability definitions have to expose representations of such definitions in the context of the specific template. In other words, while requirements and capabilities defined in a node type represent a sort of meta-data, those represented in a node template provide concrete values. Also, requirements and capabilities in a topology template can be explicitly connected by relationship templates to indicate that a specific requirement of one node is fulfilled by a specific capability provided by another node.

5.2.3 TOSCA Container: CSAR

In order to allow execution and management of a cloud application within an environment, the service template and all the relative artifacts must be available to the target platform. TOSCA defines the **cloud service archive** (CSAR), which is an archive format defined with the objective of ensuring availability of all the artifacts and templates needed to execute a certain application within a single file. Being a container file, a CSAR can contain files of multiple types, organized in hierarchical subdirectories that are specific for a particular cloud application. However, a TOSCA-meta-data subdirectory, containing a "TOSCA" meta-file, is always mandatory. The meta-file represents all meta-data relative to other files in the CSAR, in the form of name/value pairs organized in blocks, each of which provides information for a specific artifact. The first block of the meta-file provides global information about the CSAR itself, while the others always begin with a name/value pair pointing to an artifact within the CSAR, followed by properties relative to that particular artifact.

5.2.4 Implementing Tools: Winery, OpenTosca, and Vinothek

TOSCA can take advantage of some interesting implementing tools that allow to manipulate definitions of templates, types, and artifacts through graphical interfaces, to deploy them to a target platform, or to simply publish them for users. The three tools listed below are strictly related and belong to the "OpenTOSCA ecosystem."

- **Winery** [8] is a graphic environment, accessible through Web browsers, which supports modeling of both TOSCA topologies and plans. In particular, Winery provides the possibility to create and modify, through a very intuitive and easily accessible graphical interface, node and relationship types and templates. Two dedicated graphical components support topology manipulation and creation of BPMN workflows (via the BPMN4TOSCA plug-in). All information is stored in a repository that allows importing and exporting using the TOSCA packaging format (CSAR). A set of node types, comprising most of Amazon Web Services and some additional cloud available services, is already defined in Winery and is ready for the user. Figure 5.2 reports a snapshot of the Winery web-based tool. In particular, the picture portrays the graphical composition of a topology template, containing both node templates and relationship templates. On the left, the palette of predefined node types available for immediate use is clearly shown.

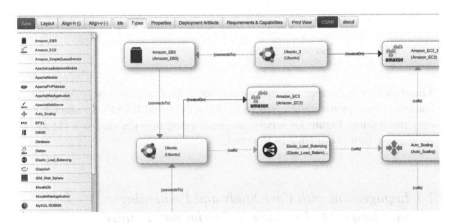

Fig. 5.2 Winery: example of topology graphical composition

- **OpenTOSCA** [9], also referred to as "TOSCA container", is an open-source browser-based runtime environment for running applications described using TOSCA specifications. In particular, while Winery can be used to graphically model TOSCA service templates to describe a cloud application, OpenTOSCA can be used to deploy such applications by importing and installing a Cloud Service Archives (CSAR), containing all files needed to instantiate the service, from templates to plans and including both implementation and deployment artifacts. Based on these archives, the management plans are used to create, operate, and manage instances of the corresponding application. Figure 5.3 shows a snapshot of the OpenTOSCA tab to import CSAR containers.

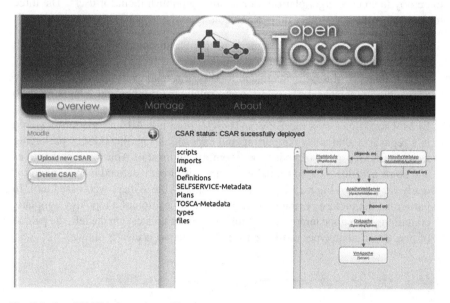

Fig. 5.3   OpenTOSCA: importing a CSAR container

- **Vinothek** [10] is another interesting tool that is used as a repository for TOSCA configurations. In particular, Vinothek offers deployed CSARs to users for an easy installation. Figure 5.4 reports the description of a CSAR for a "Moodle" application, as it is available in Vinothek.

5.2.5 Engagement with Case Study and Positioning with Respect to Use Case Scenarios and Features

TOSCA and its supporting tools, presented in Sect. 5.2, can be used to easily implement the Business Intelligence example application described in Sect. 1.2.2. In order to do so, the hypothetical owner of the Business Intelligence application catches its structure in a TOSCA Service Topology, a graph with nodes and relationships. Nodes represent the application components implemented as services and provided either as SaaS, PaaS, or IaaS, while relationships connect nodes and define the topology's structure. In particular, a possible implementation of our Business Application may contain the following nodes (Fig. 5.5):

- **SugarCRM** implements the CRM functionalities needed by the application, enabling users (whether in sales, marketing, or support) to create extraordinary customer relationships. Users can use it as a SaaS or host it as a virtual appliance on a Virtual Machine (provided by a vendor of their choice).
- **SAP ERP** represents a possible implementation of the ERP component. The chosen application, offered either as a SaaS or as a virtual appliance, is an enterprise

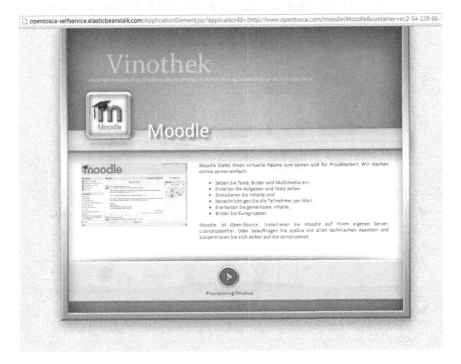

Fig. 5.4 Vinothek: CSAR for application "Moodle"

resource planning software developed by the German company SAP SE. It consists of several modules, including utilities for marketing and sales, field service, product design and development, production and inventory control, human resources, finance, and accounting.

- **Ironcluster ETL** takes care of the extraction-transformation-loading steps required by the ETL component. It takes away the complexity of data integration, delivering a much more agile ETL environment with the capacity you need. It is deployed in Amazon EMR.
- **Amazon EMR** (Elastic MapReduce) is an Amazon services that analyzes and processes vast amounts of data. It does this process by distributing the computational work across a cluster of virtual servers running in the Amazon cloud. The cluster is managed using an open-source framework called Hadoop.

- **Amazon Redshift** is a data warehouse solution that makes it simple and cost-effective to efficiently analyze all data using existing Business Intelligence tools.
- **MySQL Database** can be used to store organized collections of data, either in public or in private infrastructures. The database service can be directly offered as a SaaS or a PaaS, or be included in a virtual machine hosted on a cloud platform.

Both nodes and relationships are typed entities and hold a set of type-specific capacities, giving a subject and variability to generic TOSCA elements. Each node

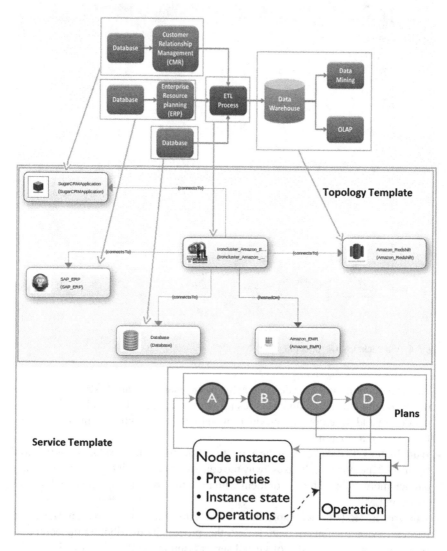

Fig. 5.5 Implementation of the Business Intelligence example in TOSCA

is matched with an index of operations it offers to manage itself. One key to support interoperability and reusability is that nodes expose their management operations explicitly as part of the topology. Management plans combine these management operations to create higher-level management tasks, which can then be fully automated and performed to deploy, configure, manage, and operate the application. Management plans are the key to access the interoperability capabilities offered by TOSCA. In particular:

- Plans orchestrate the management interfaces and operations defined in TOSCA nodes. Operations can be described using the Web Services Description Language (WSDL), through Representational State Transfer (REST) APIS, or by means of scripts that implement specific management operations on the respective node. These operations might be external services, or their implementation might be included in the service template as an implementation artifact: in the latter case, the management environment guarantees that such implementations are settled before the service template is instantiated. The orchestration provided by plans fully supports services' interoperability at all levels.
- Plans can inspect the topology model to access nodes and relationships used to describe a service or set of services: in this way, each change affecting the topology is reflected on the orchestrating plan, which flexibly adapts itself to the new configuration. In this way, if one or more nodes were to be substituted or modified, for example in order to replace a service offered by a provider with another one hosted on a different platform, the application's deployment would be still possible without major modifications. Application portability surely benefits from such a feature.
- Plans read and write a service instance information (the nodes instance state, such as properties containing credentials, IP addresses, and so on). The workflow engine manages the state inside a plan and releases it to the different activities.

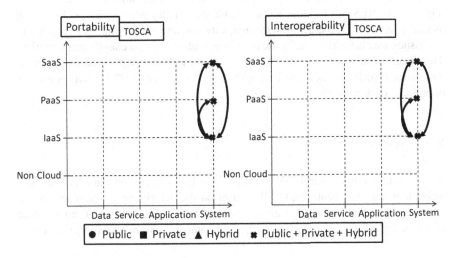

Fig. 5.6 TOSCA approach positioning

TOSCA plans' portability derives from the workflow language and engines used to describe and run them. Workflow languages, such as BPMN [7] and BPEL [6], are generally portable between different engines. In [11, 12], it is possible to find further considerations of TOSCA's ability to support both the portability and interoperability of cloud applications. In particular, the positioning of the TOSCA standard according to our n-dimensional space is illustrated in Fig. 5.6. The TOSCA model is flexible enough to be seamlessly applied to different use cases at all service levels. The implementation of our test application demonstrates the TOSCA capability to support the migration of a whole software system to the cloud (CSCC5). Considering that the TOSCA model can be used to represent in-house capabilities, CSCC4 can also be realized. An application that has been ported to the cloud by means of TOSCA, can be composed of services provided by different cloud providers (CSCC2) and a user can decide to substitute a specific service with a similar one from a different vendor (CSCC1).

5.3 Cloud Infrastructure Management Interface

Cloud Infrastructure Management Interface (CIMI) [13] is a standard proposed by the Distributed Management Task Force (DMTF) which specifies an interface, represented by a set of RESTful APIs, to manage cloud platforms operating at an infrastructure as a service layer. The specification documents focus on the description of a RESTful interface, but the standard separates the API design from the particular communication protocols to use: in particular, future developments will include SOAP and RPC implementations of the same interface. The interface mostly focuses on the management of the IaaS infrastructure, not on the services such infrastructure offers. The CIMI model defines a set of resources, associated templates, and configurations, which can be accessed, operated, and managed through the basic HTTP methods in a RESTful fashion. These include cloud entry points to access lists of all available assets, virtual machines, storage, network, and monitoring resources. Security issues are also addressed by the interface, with a focus on client's identification. The documentation offers a precise description of all the steps required to access and manage such resources, including the exact sequence of HTTP calls needed to operate on each resource.

5.3.1 Scope

It is important to understand CIMI's scope, in order to determine when and how it is possible to use it. Essentially, CIMI focuses on the description of the management interface of a cloud infrastructure, but it does not go beyond that. The consumer of an IaaS platform can create, wire up, and control the infrastructure for an entire system, and even instantiate virtual machines with software preinstalled on them, but all the following interactions with such software could not be handled through CIMI. The consumer must use other means to install and manage complex services implemented in an IaaS cloud: in this case, other standards like TOSCA can be extremely useful. Since CIMI focuses on the management interface only, not all kinds of resources in an IT infrastructure are modeled: for example, operative systems are not represented even if they are generally preinstalled on a virtual machine (which is instead manageable through CIMI). The CIMI specification focuses on a REST-style protocol but, thanks to its design which separates the management interface from the communication protocol, different solutions will be available in the future.

5.3.2 CIMI Model

The core of the CIMI interface is represented by a set of resources used to represent the IaaS components. In particular, the standard describes two sets of elements: infrastructure components like machines, storage volumes, and networks; monitoring-related concepts and artifacts, like events, meters, and logs. Each resource description consists of three main elements: a template resource, a configuration resource, and the resource itself. A **template** represents a description of a resource with a preexisting configuration that can be used as the basis to define new resources or can be directly instantiated. Providers could offer catalogs of preconfigured templates which consumers could choose from, or let users define the desired resource from scratch. A typical template contains information like CPU type and speed, available memory, disk space, and network configurations. **Configuration resources** are supplied with templates and are used when a consumer wants to modify the default template with his own customizations. For instance, if a user wants to add more CPU power to an existing template, he can do that through a configuration resource associated to that specific template. The combination of configuration and template resources composes the final service the provider will instantiate and offer to the client. CIMI **resources** represent the infrastructural element of the IaaS platform at the disposal of a consumer, which are modeled as fully accessible REST resources: a user can reach them through URIs, found in their description, and operate on them using the basic HTTP methods. A list of the available resources on a provider's cloud is accessible through a **cloud entry point**, which can be used by providers to hide or show resources according to the users' privileges and roles. When defining a resource the CIMI standard describes a usage pattern common to almost all the elements of the specification, which includes all the steps and messages (with HTTP examples) users have to exchange with the platform in order to instantiate that particular resource. When a user wants to instantiate a resource, he needs to gain access to the cloud entry point, the address of which should have been previously provided by the vendor. Registration and login policies can be enforced by the specific vendor, but the standard does not provide support to this. Once the user has access to the entry point, she can retrieve a list of available templates for a specific resource, or issue a command to instantiate it without a default configuration. The resources described by the CIMI standard include the main infrastructural elements needed to configure and manage an IaaS cloud and they consist of:

- **Machines** define single computer systems usually referring to virtual machines, but sometimes also describing physical hardware according to the provider's needs. The user, once inside the Cloud Entry Point, can act in two ways: she can retrieve a list of machine images and configurations to launch and instantiate; alternatively, she can request the instantiation of a template, if available.
- **Volumes** represent storage resources. Their instantiation is very similar to that of machines but, in order to access and use them, they need to be associated to an existing machine or network.
- A **Network** represents an abstraction of a transportation network, consisting of interfaces, ports, and links, but its definition is not related to a physical realization.
- **Jobs** are monitoring resources related to CIMI operations: a consumer can query jobs to determine the state of an operation. While it is not mandatory for providers to define job resources, if they decide to do so they need to expose a job for every operation able to alter the state of a component inside the system.
- **Meters** are used to monitor the health and performance of the system, as they represent, for a certain resource, the specific property to check and how to monitor it.
- **Event** resources are directly created by providers and not by the client, who can only consult event logs through the cloud entry point. They represent information tracked by the provider and exposed to the consumer, consisting of a time stamp, a type, contents, a severity indicator, and an outcome of the event.
- **System** and **system template** recall concepts represented by the OVF system and the TOSCA service template, as they represent groups of resources designed to work together to provide a certain service or to accomplish a set of tasks. System templates can be considered as patterns that describe the infrastructure of a complex service, composed of multiple machines, networks, routing groups, storage, and monitoring resources. Each template can be used to replicate the same system multiple times.

5.3.3 Security

The CIMI specifications indicate which points of the management interface must be secured and suggest other areas where security may be necessary, without enforcing it. In particular, the CIMI security model defines two security domains: API security, dealing with the management interface and in full scope of CIMI; resource security, related to safety of instantiated resources and thus only partially covered by the specification.

- **Resource security** relates, according to the CIMI model, to all resources running on a cloud platform using the CIMI interface. While security of the management interface is handled by the CIMI specification, ensuring only authorized accesses to management functionalities, functional interfaces of instantiated resources are the consumer's responsibility. As mentioned before, CIMI introduces the *Credential* resource in order to identify clients who instantiate a system component (a machine, a volume, and so on), but everything else is left to the user.

- When considering **API security**, the CIMI specification focuses on five security areas, namely that are: authentication, message integrity, message confidentiality, authorization, and multi-tenancy. Of these, particular focus is on authentication and authorization. The consumer of a system must always be securely identified in order to access the management interface. Also, the available resources and relative operations are exposed to the client according to her permissions and privileges. Use of secure protocols to exchange messages is encouraged, but not imposed: for example, use of the HTTPS protocol is suggested, but the specification does not require providers to implement it nor directly supports its use.

5.4 Cloud Data Management Interface

Cloud Data Management Interface (CDMI) [14] is a standard for managing data on cloud platforms, proposed by the Storage Networking Industry Association (SNIA). CDMI defines a functional interface that users and applications can use to create, retrieve, update, and delete data elements from cloud storages. Using the interface, clients can also discover the capabilities offered by the cloud platform and manage the containers and the data that is placed in them, together with meta-data associated to both containers and data. Other capabilities supported by the standard are the creation of queries to retrieve data, management of user permissions and groups, access control, queue usage, and so on. Administrative and management applications can leverage these capabilities to manage containers, accounts, security accesses, and monitoring/billing information. All the elements needed to describe the data store offering, which includes containers and meta-data, are fully described in the standard, together with APIs used to access them. The CDMI interface is RESTful and, currently, no other implementations are included in the standard nor have plans been made to do so in the near future. Unlike other standard interfaces using both XML and JSON to code messages, CDMI only supports JSON in order to reduce payload dimensions.

5.4.1 Core Concepts

The standard defines a set of components which are used to represent entities in a cloud storage. Figure 5.7 reports the main elements defined in the CDMI model and their relations.

Among the different entities defined by the standard, the concept of **capability** is central for the exact description of a cloud storage's characteristics. Capabilities

describe the functionality implemented by a CDMI server and are used by a client
to discover supported functionality, represented through a set of configuration parameters. The usefulness of a capability description is clear: a client who wants to use
a certain functionality just has to check if the target cloud storage exposes

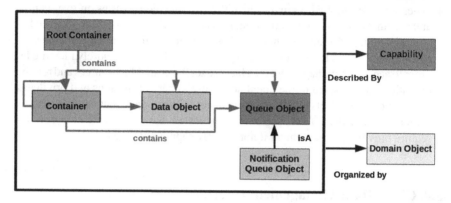

Fig. 5.7 Overview of the object defined in the CDMI model

that particular capability. The CDMI standard describes cloud storages through a
filesystem-like structure, in which **Data Objects** contain the stored data and are
organized as files in a hierarchical directory, where folders are instead represented
by **Containers**. A container can virtually contain any number of data objects, but
limitations may be imposed by the cloud provider. CDMI also defines resources to
manage data accesses and user/group policies. In particular, the standard supports
the concept of **Domain**. Domain objects define a logical grouping of objects that are
meant to be managed together or by a specific set of users, enabling an administrative
management of the cloud storage. They can also be hierarchically organized, so
that parent and child domains can exist to better delimit objects' ownership. Users
and groups belonging to the same domain can refer to each other directly, without
reference to any other domain or system. Domains also work as containers for usage
and billing summary data: measures about objects' use, associated to each domain,
produce information that can be passed from child to parent, thus simplifying billing
and management operations that are typical of a cloud storage environment.

5.4.2 Queue Objects

CDMI models' **Queue Objects** are used to store zero or more values and are accessed
in a first-in-first-out manner. POST HTTP messages are used by object writers to put
data in a queue, while the reader uses GET messages to retrieve value(s) from the
queue object and subsequently deletes them to acknowledge their receipt. Queues

provide a simple but effective mechanism for one or more writers to send data to a single reader in a reliable way. CDMI defines **Notification Queue Objects**, used to allow clients to discover changes in the system efficiently. The standard supports persistent queues, which can be between clients or are specific for certain users who can operate on them independently. The discriminating factor between simple queues and notification queues is represented by the attached meta-data. When defining a notification queue, clients define a standard one and then add meta-data with instructions regarding the type of queue to instantiate and the notifications they want to receive, with the information to be attached. Clients can also create notification queues even if the CDMI server does not expose the relative capability: simply, no notification will ever be put in that queue, which will be treated as a standard one.

5.4.3 Security

The CDMI specification addresses different protective measures to apply in order to manage and access data and storages safely. However, not all security measures are mandatory in the CDMI standard: before exploiting a CDMI server, a user should verify which security methods that specific server applies by analyzing its capabilities description and only then should he take a risk-based decision on whether to use it or not. This is particularly important if objects to be stored on the server contain sensitive or regulated data, requiring some particular protection (encryption, for instance) or the use of which needs to be accurately tracked. CDMI considers several security aspects:

- Protection of communications between CDMI clients and servers (confidentiality and integrity of messages).
- Mutual identification of CDMI clients and servers.
- Restrictions on CDMI clients' permissions and actions based on their domain and privileges.
- Auditing and tracking of actions performed by a client on a server.
- Protection of stored data, not currently used by a client (data at rest).
- Strong control on data elimination to prevent loss due to malicious interventions.
- Discovery of a CDMI server's security capability.

CDMI servers are obliged to offer transport security and to expose their security capabilities to users, while other measures may vary from one implementation to another. HTTP is the mandatory transport mechanism, and HTTP over TLS (i.e., HTTPS) is the mechanism used to secure the communications between CDMI clients and servers. To ensure both security and interoperability, all CDMI implementations implement the Transport Layer Security (TLS) protocol, but its use by CDMI clients and servers is optional. As for access control, CDMI follows the ACL and ACE model used for file authorization operations by NFSv4.

5.5 Open Cloud Computing Interface

The Open Cloud Computing Interface (OCCI) [15] is a RESTful protocol and API, published by the Open Grid Forum (OGF), as a result of a community effort. The objective of the proposed standard is to define a shareable and homogeneous interface to support all kinds of management tasks in the cloud environment. While the original scope of OCCI covered the creation of a remote management API for IaaS platforms, at the moment the proposed interface is suitable to represent other cloud models, such as PaaS and SaaS, but it could also be applied to other programming paradigms. The OCCI specification is released as a set of complementary documents, classifiable under three categories:

- **OCCI core specification** consists of a single document defining the whole OCCI core model. More on this model will be presented in the following sections. The core model can be expanded and interacted with.
- **OCCI rendering specifications** comprise several documents, each of which describes a specific interaction model to communicate with the OCCI Core. Multiple renderings can interact with the same instance of the OCCI core model and will automatically support any additions to the model which follow the extension rules defined in OCCI core. The OGF currently proposes a rendering for HTTP, defining how to communicate with and serialize the Core, using the HTTP protocol.
- **OCCI extension specifications** describe extensions to the OCCI core model, introducing additions to the OCCI core model defined within the OCCI specification suite. The OGF has proposed an IaaS extension with specific resource types described through sets of operations, attributes, and relationships.

5.5.1 The OCCI Core Model

The OCCI core document describes the standard model at a high level of abstraction: in this way, no specific application domain is ever addressed and no limitation is given to its use. Furthermore, the specification itself claims its applicability to either the PaaS, SaaS, or IaaS. The model defines a hierarchical set of elements, which can be further extended and specialized through OCCI extension specifications. Figure 5.8, extracted from the OCCI core specification available at [16] reports a UML Class Diagram describing the elements of the model. The root element of the core model is represented by **Entity**, an abstract class containing the base characteristics shared by all other components of the specification, like properties such as ID, name, and inclusion in **Mixin** instances (more on this below). Being abstract, it cannot be directly instantiated.

The concrete classes deriving from Entity are represented by:

- **Resource** represents the focus of the specification. In order to be OCCI compliant, the resource class has to implement all the attributes from the Entity class, plus

other properties defining a summarizing description of the resource and a set of
Links to other resources.

- **Kind** represents a repository of type-specific information. A resource's or link's
  properties are not directly defined in their description: the properties unique to

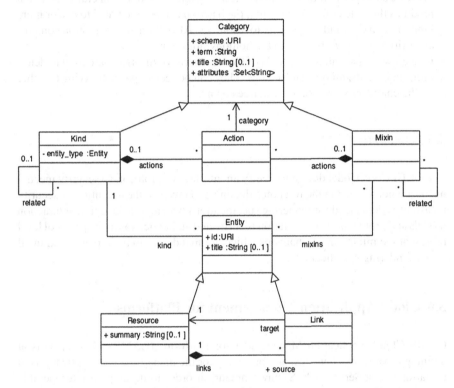

Fig. 5.8   UML class diagram of the OCCI core model

a descendant of Entity (either a resource or a link) belong to the kind class it is associated with. Each resource must be associated with one and only one kind instance.

- **Link** represents the base relationship existing between two resources. Deriving
  from Entity, it inherits its basic properties but also specifies a *target* and *source*
  attribute referring to resource instances within the service provider's name-space.
- **Mixin** instances are used to add functionalities to resources, even at runtime, but
  they cannot be applied to kinds. They slightly resemble Java interfaces, since they
  expose sets of operations and attributes an associated resource can leverage, just
  by declaring its participation to a mixin. However, while Java interfaces contain
  blank methods that the inheriting class must override with its own code, a mixin
  class already contains the implementation of its operations, which can be immediately used by the participating resource. The resource has no obligation toward

the mixins it participates in: it can just use the provided functionalities with no further attachments.

- **Action** represents a specific operation that may be performed on an Entity's descendant or on an entire collection thereof. The action class is abstract, so it is necessary to create subtypes of it in order to define a concrete operation. Generally, an action modifies the state of the target entity. Each action can be associated to one or more kinds and mixins, and not directly to resources. Also, the action definition contains an attribute *category* tying it to the homonymous class.
- **Category** is associated to Entities (either resources or links) and are used to determine their kinds, mixins, and actions. They can be seen as packages tying together all the characteristics a set of resources owns.

5.5.2 Security

The OCCI Core Model does not specify an interface or protocol, so security mechanisms are not defined in the reference document. However, the definition of security measures is demanded for rendering specifications which, instead, define interaction models and protocols. Transportation security and authentication at the protocol level represent the minimum mechanisms which are mandatory to be implemented in all OCCI rendering specifications.

5.6 Cloud Application Management for Platforms

OASIS Cloud Application Management for Platforms (CAMP) TC [17] aims at defining models, mechanisms, and protocols for the management of applications in a Platform as a Service (PaaS) environment, in order to develop an interoperable protocol for PaaS management interfaces that users can exploit to build, deploy, and administer their applications. CAMP's goal is to define a simple standard RESTful API, along with a JSON-based protocol, with an extensibility framework that enables interoperability across multiple vendors' offerings. The current documentation contains a set of basic APIs a cloud vendor should provide as part of its PaaS offer, in order to manage the building, running, administration, monitoring, and patching of applications. Also, a resources model is provided to describe the main components of any platform offer. This would enable interoperability among self-service interfaces to PaaS clouds, through the definition of artifacts and formats shared between conforming cloud platforms. Also, this would allow independent vendors to create tools and services that communicate with any CAMP-conform cloud platform, using the defined interfaces. Vice versa, cloud vendors could easily exploit these interfaces to develop new PaaS offerings, or adapt the existing ones, which would be compliant with independent tools. The current specification already supports a series of possible use cases, which include the possibility to run, stop, patch, and restart applications, or to build applications in a local environment and then run them in the cloud.

5.6.1 CAMP Model

Being based on a RESTful API, the CAMP standard defines all of its components as resources accessible through URIs, on which different commands can be issued as POST/GET HTTP calls. Messages exchanged through HTTP are encoded via JSON strings. As for other standards, resources share a basic set of properties, all defined in a root element named **camp_resource** acting as parent class for all other entities. There is no clear hierarchical organization between resources, since they all seem to directly inherit from the base camp_resource class. CAMP specification, available at [18], contains UML class diagrams accurately describing its model and components. Resource components in the model are represented as follows:

- **Platform** reports a view of the running platform, comprising all the resources running on it in every instant. It exposes collections of resources representing the services provided by the described platform (service resources), applications currently running on it (assemblies), and meta-data defining supported resources and extensions.
- **Assembly** is a resource representing applications running on the reference platform. All operations executed on an assembly resource are reflected on components and elements of the relative application.
- **Component** resources can be part of one or more assemblies as they represent discrete and reusable elements of an application. An assembly is composed of one or more components. Different relationships between components can be specified.
- **Plan** resources are meta-data providing information on all the elements composing applications, including artifacts, services required to execute and utilize such artifacts, and relationships existing between them. Artifacts represent static components of an application, whereas *Component* resources generally define dynamic elements. In order to describe a plan, CAMP provides the possibility to represent it as a resource or through a YAML [19] file. *Service* resources represent functionalities exposed by the reference platform as services, which can be used to operate on, create, and destroy platform components.

5.6.2 Operations and Sensors

The resources described so far are not the only ones present in the model. In order to interact with an application deployed in a CAMP-compliant PaaS platform via the CAMP API, *Operations* and *Sensors* resources have been defined.

An **Operation**, also known as "effector," represents any action that can be taken on a resource. A **Sensor**, on the other hand, represents data about a resource which can dynamically and rapidly change during the resource's lifetime, or that needs to be accessed from external systems. Measures taken on the resource according to predefined metrics or the very resource's state represent a good example of sensors. Assemblies and components can expose multiple operations and sensors, which enable both consumers and providers to manage them through the CAMP API.

5.6.3 Application Deployment

Deploying an application using the CAMP API is just a matter of sending an HTTP POST message containing information about the desired application or having some configuration files attached. To ease deployment and support migration across multiple platforms, CAMP defines the **Platform Deployment Package** (PDP): a simple archive of executable images, dependency descriptions, and meta-data that can be used to move an application and all of its components from one platform to another, or between a development environment and an operative target platform. The archive generally contains a plan file (expressed in YAML), together with application content files such as web archives, database schemas, scripts, source code, localization bundles, and icons. In the simplest scenario, either a single plan file or a PDP can be used to create an assembly resource through an HTTP POST request. A platform supporting plans as CAMP resources provides consumers the means to build a plan element from a PDP or a plan file. Multiple assembly resources can be created from a single plan resource by submitting multiple HTTP POST requests.

5.7 Cloud Standards Coordination Initiative

Following a specific request from the European Commission, the European Telecommunications Standards Institute (ETSI) [20] launched the Cloud Standards Coordination (CSC) [21] initiative. The main objective of this initiative, which submitted its final report in November 2013, was to collect and analyze the current standards and technologies, applicable to the cloud computing paradigm, and to determine how each of them could address one or more specific cloud issues. The targeted audience is wide, as it includes:

- **Cloud service providers** of all types and dimensions, who can use the final report as a "compass" to choose the best formalism to use when describing their services.
- **Cloud service customers** (including governmental entities), who should be able to clearly understand if the offered services meet their requirements, thanks to a transparent and standard description of them.
- **Governmental authorities** acting as cloud regulators.

The final report offers a deep analysis of the current state of the art regarding cloud standards, together with the definition of various roles in cloud computing and an interesting number of use cases. Organizations involved in cloud computing standardization are taken in great consideration, and a selection of their documents, reports, and white papers has been referenced, together with a mapping between the standards provided and the activities involved in the whole cloud service lifecycle (which are also classified). Interoperability and portability are addressed specifically in a dedicated paragraph "The Interoperability Perspective," where the issues are briefly introduced and discussed. Interoperability is further investigated in one of the use cases presented in the report, "Cloud Bursting," described below.

5.7.1 Role Definitions

The CSC initiative provides a high-level taxonomy of cloud stakeholders, organizations, or individuals acting either as providers or consumers of cloud services. Such taxonomy, reported in Fig. 5.9, comprises a set of role definitions:

- **Cloud Service Customer** identifies each subject consuming one or more cloud services exposed by a provider.
- **Cloud Service Provider** exposes services which are then consumed by providers.
- **Cloud Service Partner** supports the provisioning of services offered by a provider or their consumption by a customer.
- **Government Authority** interacts with customers, providers, or partners to enforce laws and regulations.

Also, the report defines the concept of **Party**, representing an individual or organization that can play different roles at the same time.

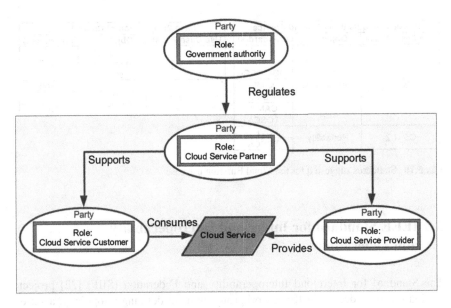

Fig. 5.9 CCS defined roles

5.7.2 Use Case Descriptions

The initiative proposes a number of interesting use cases involving the roles we have just presented. For each use case, it also reports the standards which can be used to implement a certain step. Each use case follows three well-defined steps, which are specialized for the particular scenery described.

1. **Acquisition of Cloud Service**: This step describes all the preparatives necessary to use a set of cloud services from different providers. These may include SLA negotiations, agreements on data formats or on standards for API interfaces for interoperability and portability purposes.
2. **Operation of Cloud Service**: In this phase, the activities characterizing the use case are described. In particular, the scenario reports how the different services communicate and in which order, together with the difficulties and challenges which could be encountered.
3. **Termination of Cloud Service**: In this last phase, all the activities connected to the termination of a service or a set thereof are described. These include the operations needed to stop running virtual machines or to log out users, to free IP addresses or shut down a router, and so on.

Also, for each step in each use case a list of useful standards is included: for the *Cloud Bursting* use case, reported in Sect. 1.2.1, the different standards proposed are shown in Fig. 5.10.

Phase 1: Acquisition of Cloud Service				Phase 2: Operation of Cloud Service		
Use Case Phase	Objective	Standard(s)		Use Case Phase	Objective	Standard(s)
CB_1.1	Interoperability	OGF/OCCI		CB_2.1	Creation of a VM for public cloud	OVF
		CIMI		CB_2.2	Provision of the infrastructure	CIMI
		CDMI				OCCI
		CAMP				OVF
		TOSCA				
CB_1.2	Portability	OVF				
		TOSCA				

Fig. 5.10 Standards suggested for the Cloud Bursting use case

5.8 IEEE Standard for Intercloud Interoperability and Federation

The Standard for Intercloud Interoperability and Federation (SIIF) [22] project, carried out by a dedicated IEEE workgroup, aims at defining a topology, a set of functionalities, and a governance model to support cloud interoperability and federation between different platforms. The scope of the resulting architecture is to ease the building of intercloud solutions, enabling communication between different platforms thanks to a shared set of standards and resources definitions. Description of internal cloud behavior (intracloud) is completely disregarded and out of scope. The current standard, still in development and only reported as a draft, puts a strong focus on the description of the intercloud topology, which makes reference to the NIST definition of cloud computing [23], defining in detail its components and the relationships between them. Each element of the topology has a clear and fundamental role in the resulting architecture. Different standards and protocols, not exclusively related to the cloud, are taken into consideration to enable communication between the described components. Different scenarios are used to describe the basic functionalities offered by the topology and the adopted protocols.

5.8.1 The Intercloud Topology

The intercloud topology is composed of a set of well-defined components, with peculiar characteristics and dedicated protocols. Among the elements defining the topology, we identify:

- **Intercloud Root** provides a set of capabilities comprising the management of the Cloud Resources Directory Services (used to expose resources in a shared format), the Trust Authority Service (for SLA and policy application), and Presence Information. Each root is connected to an instance of intercloud exchanges, acting as a mediator for enabling connectivity between multiple cloud environments. The Cloud Computing Resource Catalog hosted at the roots is a holistic and abstract view of the computing resources available across several cloud environments. Individual platforms use this catalog to identify resources matching with a set of preferences and constraints. The root is not a central repository: intercloud roots will host the globally dispersed computing resource catalog in a federated manner, scaling and replicating P2P technologies.
- **Intercloud Exchanges** support the resources' negotiation and collaboration between heterogeneous cloud environments, by leveraging the distributed catalog hosted by intercloud roots in order to match cloud resources, by applying preferences and constraints expressed by consumers. Processing nodes are organized according to a peer-to-peer model, based on a DHT overlay approach in order to facilitate optimized resources matchmaking queries. Ontology information is replicated to the different DHT overlay nodes from their affiliated intercloud roots using a "Hash" function. Intercloud exchanges also play a key role in security and trust: during the identification process, consumers specify a "Trust Zone", whose exchange nodes are included in the matchmaking constraints.
- **Intercloud Capable Individual Clouds** communicate with each other through the environment created by Roots and Exchange: in this communication, they act as a client, while the overall environment represents a server. Connections to the Intercloud are possible through intercloud gateways only. Once the initial negotiating process is completed, communications between cloud instances are managed directly by the participants, through a shared set of protocols and standards.
- **Intercloud Gateways** represent the access points to the Intercloud environment, providing mechanisms to support the adopted protocols.

More information and details about the Intercloud architecture and considerations on security issues in a federated cloud environment can be found in [24, 25].

5.8.2 The Intercloud Protocols and Standards

The intercloud project has chosen a set of protocols and standards, born outside of the cloud scenery, in order to establish a common formalism for communications between trusted federated cloud providers. The objective is to provide ubiquitous and interoperable content, storage and computing capabilities in a network of clouds. The base protocol adopted by Intercloud is represented by the **Extensible Messaging and Presence Protocol** (XMPP) [26], representing a viable control plane presence and dialog protocol. XMPP root services will be located in the intercloud roots. XMPP defines protocols for communicating between groups of entities which register with an XMPP server: in the intercloud vision, multiple XMPP servers are connected together, providing dynamic registration capabilities. XMPP facilitates both presence and many-to-many messaging across service provider domains. XMPP messages are extensible and can be used to carry messages of different types: a specific extension for cloud will be used. Security in communications over XMPP is managed through specific protocols. In particular, XMPP supports encryption based on the **Transport Layer Security** (TLS) **protocol** [27], along with a "STARTTLS" extension that is modeled after similar extensions for the IMAP and POP3 protocols, **Simple Authentication and Security Layer** (SASL) [28] and **security assertion markup language** (SAML) [29] are used to provide secure authentication in a federated environment. In order to support Service Discovery and resources' sharing, XMPP-based RDF and SPARQL approaches are investigated. Remarkable is the representation of the cloud resources, which is carried out by means of semantic-based standards such as **OWL** [30] through the "mOSAIC Cloud Ontology," produced during the mOSAIC project, which aims at developing an open-source platform that enables applications to negotiate cloud services as requested by users. The mOSAIC project is part of the EU FP7-ICT program.

5.9 Intercloud Architecture for Interoperability and Integration

The Intercloud Architecture Framework (ICA) [31] represents the focus of an ongoing research project carried out by the System and Network Engineering (SNE) group [32], which addresses integration and interoperability issues in multi-provider and multi-domain heterogeneous cloud-based infrastructures, also taking into account legacy infrastructure services. Current documentation, available at [33] at the time of writing, takes in great consideration the existing standards and tries to extend previous works in order to build a homogeneous model for cloud computing. In particular, the specification focuses on the NIST Cloud Computing Reference Architecture (CCRA) in order to build ICA, but it is also influenced by works proposed by the IEEE Intercloud Working Group (see Sect. 5.8) and ITU-T Focus Group on Cloud Computing (FG Cloud) [34].

5.9.1 Scope of the Work

As stated earlier, the proposed intercloud architecture focuses on interoperability and integration issues in multi-domain and multi-provider cloud environments, without forgetting interactions between cloud and legacy platforms. Several issues are taken into consideration when facing interoperability and integration problems:

- Communication between applications and services represents the first crucial point of discussion. The ICA model considers interaction between cloud applications and services across the different service layers, also "vertical integration," and between cloud domains and heterogeneous platforms, in which case we speak of "horizontal integration." Compatibility of the model with the different cloud service models (IaaS, PaaS, and SaaS) is necessary to provide integration.
- ICA also addresses the possibility to enable applications to directly control infrastructure resources and services at different layers, in order to optimize their organization and use. This includes the definition of a common **intercloud control plane** and of signals for cloud services and network integration.
- Another important point of discussion is the automatic provisioning of services and infrastructure resources, together with their lifecycle management, including deployment, monitoring, and composition. Services and resources from multiple providers should be treated homogeneously in order to reduce interoperability issues.
- ICA aims at supporting intercloud federation through a well-defined and shared framework, which could explicitly model and support instantiation of intra- and intercloud networks.

5.9.2 Elements of the Framework

The ICA Framework proposes a set of complimentary components which, used together, would allow to completely model a federated cloud environment by answering to the different key points listed previously.

The multilayer **Cloud Service Model** (CSM) component aims to homogenize and integrate the different cloud service models. The scope of CSM includes communication between the different layers, definitions of links among their components, and creation of common interfaces between different layers. Using an ISO/OSI pile style, the CSM defines a set of layers organized from top to bottom as follows:

- C6: User/customer side resources and services
- C5: Access/delivery infrastructure hosting components and functions to provide access to cloud services/resources and interconnect multiple cloud domains

- C4: Cloud services layer that may include different types of cloud services: IaaS, PaaS, SaaS
- C3: Cloud virtual resources composition and orchestration layer that is represented by the cloud management software (such as OpenNebula, OpenStack, or others)
- C2: Cloud virtualization layer (e.g., represented by VMware, Xen, or KVM as virtualization platforms)
- C1: Physical platform (PC hardware, network, and network infrastructure).

Intercloud Control and Management Plane (ICCMP) supports interactions between different cloud applications by providing signaling, synchronization points, session management, and infrastructure optimization functionalities. **Intercloud Federation Framework** (ICFF) represents the core component for enabling federation of cloud entities independently managed by different providers or belonging to separate cloud domains. Such entities include services, applications, name-spaces, and even semantics. Use cases and a set of actors/roles are also defined in the specification when describing the ICFF. In particular, for the definition of actors, the specification follows the **Resource Ownership Role Action** (RORA) model proposed in the GEYSERS European project [35]. Actors defined include the following:

- **Cloud Service Provider** (CSP) represents the entity providing cloud services to customers, based on their explicit request and respecting agreements expressed through service level agreements (SLAs).
- **Cloud Broker** is a particular actor that does not provide or consume cloud services, but are in charge of discovering them according to consumers' requests and offering cloud services. Negotiations between many CSPs or customers and management of services from multiple providers can be functionalities offered by a broker.
- **Customer** is the actor that requests one or more cloud services. In the simplest case, a customer can be a single user requesting and consuming a service. In general, she can also represent an entire organization requesting a service for all its members to use.
- **User** is the final consumer of a cloud service. While a customer makes a request for a service, the user actually exploits it. In the general case, a customer represents an organization, its members are all potential users of the requested service.

Intercloud Operation Framework (ICOF) includes functionalities to support operations involving multi-provider infrastructures, such as the definition of business workflow, SLA management, and accounting. ICOF defines roles, actors, and relationships between them in terms of resources operation, management, and ownership. **Intercloud Security Framework** (ICSF) offers security mechanisms for the protection of all cloud components operating in the intercloud federated environment. In particular, it owns capabilities for integration of security measures exposed by the different layers of the CSM component.

5.10 De Facto Standards in Cloud Computing

A "de facto standard," also a market-driven standard, is a widely accepted and adopted standard which has not been defined or endorsed by industry groups (such as the W3 Consortium) or standards organizations (such as ISO). These standards can arise either because a high number of users like them well enough to collectively adopt them, or just due to their imposition on the market by some already well-established companies. Market-driven standards can become de jure standards if they are approved through a formal standards organization. As regards the cloud IaaS offer, Amazon has been for a long time the market leader with **Amazon Web Services** (AWS) and its position as one of the world's leading options for cloud-based data storage and data warehousing is beyond discussion. This is why many see AWS as the de facto standard in the public cloud. Their API is highly proven and widely used, their cloud is highly scalable, and they have by far the biggest traction of any cloud. The open-source counterbalance to Amazon's dominance is surely represented by **OpenStack**. Managed by the OpenStack foundation, it is released under the Apache license and can count on the support of a consistent developers' community. It also receives a lot of support from large IT companies including Oracle, IBM, Red Hat, and RackSpace. These companies now include OpenStack-compliant solutions into their cloud offerings or they are starting to build their products completely around this open platform (IBM efforts are surely remarkable), thus trying to impose OpenStack as the future de facto standard for IaaS platforms. For what concerns the PaaS world, the number of solutions is steadily rising. Initially, the major players were **Microsoft Azure** and **Google App Engine**, which are still regarded as the main actors in the PaaS scenario. For this reason, their solutions are widely adopted. However, recently other PaaS platforms that could be candidates to become a de facto standard have emerged. Among these are worthy of attention **OpenShift** and **Cloud Foundry** [36]. Most of the mentioned platforms are described in Chap. 4.

References

1. European Commission: Digital agenda for Europe. http://ec.europa.eu/digital-agenda/
2. European Commission: The European cloud computing strategy. https://ec.europa.eu/digital-agenda/en/european-cloud-computing-strategy
3. Lutz, S., Keith, J., Burkhard, N.L., Tsakali, M.: The future of cloud computing. http://cordis.europa.eu/fp7/ict/ssai/docs/cloud-report-final.pdf
4. Lutz, S., Keith, J., Burkhard, N.L., Tsakali, M.: A roadmap for advanced cloud technologies under h2020. http://cordis.europa.eu/fp7/ict/ssai/docs/cloud-expert-group/roadmap-dec2012-vfinal.pdf
5. Topology and Orchestration Specification for Cloud Applications (TOSCA). Vers. 1.0. OASIS Standard (2013)
6. Jordan, D., Evdemon, J., Alves, A., Arkin, A., Askary, S., Barreto, C., Bloch, B., Curbera, F., Ford, M., Goland, Y. et al.: Web services business process execution language version 2.0. OASIS Standard **11** (2007)

7. Business Process Model and Notation (BPMN). Object Management Group, Inc. (OMG) (2011)
8. Kopp, O., Binz, T., Breitenbücher, U., Leymann, F.: Winery—a modeling tool for Tosca-based cloud applications. In: Service-Oriented Computing, pp. 700–704. Springer (2013)
9. Binz, T., Breitenbücher, U., Haupt, F., Kopp, O., Leymann, F., Nowak, A., Wagner, S.: OpenTOSCA—a runtime for TOSCA—based cloud applications. In: Service-Oriented Computing, pp. 692–695. Springer (2013)
10. Breitenbücher, U., Binz, T., Kopp, O., Leymann, F.: Vinothek—a self-service portal for TOSCA. In: Workshop Proceedings, p. 72 (2014)
11. Binz, T., Breitenbücher, U., Kopp, O., Leymann, F.: TOSCA: portable automated deployment and management of cloud applications. In: Advanced Web Services, pp. 527–549. Springer (2014)
12. Binz, T., Breiter, G., Leymann, F., Spatzier, T.: Portable cloud services using TOSCA. IEEE Internet Comput. 16(3), 80–85 (2012)
13. Davis, D., Pilz, G.: Cloud Infrastructure Management Interface (CIMI) model and rest interface over http. vol. DSP-0263, May (2012)
14. Cloud Data Management Interface (CDMI) Storage Networking Industry Association (SNIA) (2012)
15. Metsch, T., Edmonds, A., et al.: Open cloud computing interface-infrastructure. In: Standards Track, no. GFD-R in The Open Grid Forum Document Series, Open Cloud Computing Interface (OCCI) Working Group, Muncie (IN)(2010)
16. OCCI core specification. http://ogf.org/documents/GFD.183.pdf
17. OASIS cloud application management for platforms (CAMP) TC. https://www.oasis-open.org/committees/camp/
18. CAMP specification document. http://docs.oasis-open.org/camp/camp-spec/v1.1/camp-spec-v1.1.html
19. Ben-Kiki, O., Evans, C., Ingerson, B.: YAML aint markup language (YAML)(tm) version 1.2. YAML. org. Technical Report, September (2009)
20. European Telecommunications Standards Institute. http://www.etsi.org/
21. Cloud standards coordination initiative. http://ec.europa.eu/digital-agenda/en/news/cloud-standards-coordination-final-report
22. IEEE p2302 Working Group (Intercloud). http://grouper.ieee.org/groups/2302/
23. Mell, P., Grance, T.: The NIST Definition of Cloud Computing. Recommendations of the National Institute of Standards and Technology. Computer Security Division, NIST, Gaithersburg, MD (2011)
24. Bernstein, D., Ludvigson, E., Sankar, K., Diamond, S., Morrow, M.: Blueprint for the intercloud-protocols and formats for cloud computing interoperability. In: Fourth International Conference on Internet and Web Applications and Services, ICIW'09, pp. 328–336. IEEE (2009)
25. Bernstein, D., Vij, D.: Intercloud security considerations. In: 2010 IEEE Second International Conference on Cloud Computing Technology and Science (CloudCom), pp. 537–544. IEEE (2010)
26. Saint-Andre, P.: Extensible Messaging and Presence Protocol (XMPP): Core (2011)
27. Dierks, T.: The Transport Layer Security (TLS) protocol version 1.2 (2008)
28. Myers, J.G.: Simple Authentication and Security Layer (SASL) (1997)
29. Hallam-Baker, P.: Security assertions markup language. May 14, pp. 1–24 (2001)
30. Bechhofer, S., Van Harmelen, F., Hendler, J., Horrocks, I., McGuinness, D.L., Patel-Schneider, P.F., Stein, L.A., et al.: OWL web ontology language reference. W3C recommendation 10, 2006–01 (2004)
31. Demchenko, Y., Ngo, C., de Laat, C., Makkes, M.X., Strijkers, R.: Intercloud architecture framework for heterogeneous multi-provider cloud based infrastructure services provisioning. Int. J. Next-Gener. Comput. 4(2), (2013)
32. System and network engineering research group. Universiteit van Amsterdam. http://sne.science.uva.nl/

33. Intercloud architecture framework draft. http://staff.science.uva.nl/~demch/worksinprogress/sne2012-techreport-12-05-intercloud-architecture-draft06.pdf
34. ITU-T focus group on cloud computing (FG cloud). http://www.itu.int/en/ITU-T/focusgroups/cloud/Pages/default.aspx
35. Escalona, E., Peng, S., Nejabati, R., Simeonidou, D., Garcia-Espin, J.A., Ferrer, J., Figuerola, S., Landi, G., Ciulli, N., Jimenez, J., et al.: GEYSERS: a novel architecture for virtualization and co-provisioning of dynamic optical networks and it services. In: Future Network and Mobile Summit (FutureNetw), pp. 1–8. IEEE (2011)
36. Cloudfoundry foundation. http://cloudfoundry.org/index.html